Food and Beverage Service

FOOD AND BEVERAGE SERVICE

Anil Sagar
Deepak Gaur
Instructor of Basic Training Centre
Pusa, New Delhi

A P H PUBLISHING CORPORATION
4435-36/7, ANSARI ROAD, DARYA GANJ
NEW DELHI-110 002

Published by
S.B. Nangia
A P H Publishing Corporation
4435-36/7, Ansari Road, Daryaganj
New Delhi 110002
Ph.: 23274050
E-mail : aphbooks@gmail.com

2026

Rs. 3195/-

Printed at
Balaji Offset
Navin Shahdara, Delhi 110032

Acknowledgements

I would like to thank members of my family who have been with me all the time and big moral support. Shri K.S. Sagar, Mrs. Rita Sagar and Mr. Sunil Sagar.

The conception of this book has drawn upon a mixture of experience and study of varied literature. I would like to express my thanks to all the organisations and individuals into gave me assistance and support in the preparation of the book, in particular I would like to thank my wife, Leena Sagar who has been with me all the time and has lent tremendous during my difficult times.

Much of the information contained in this book has come from notes, fotted down by me over some years.

I would like to express my thanks to all those who have assisted and supported me in this venture.

ANIL SAGAR

Preface

Numerous books have been published to guide and familiarise the students with the hotel industry and the service part of it. The student gets a fair idea of things as a whole. I have tried to compile the basic functions which are very important in career building process. My aim is to make the reader understand the workflow, day to day functioning of the restaurant operations. With a long experience of working with numerous five star hotels, I got an idea that the new comer in the industry at times feels stranded when in the operation. All books emphasise on the theory part only. In this book the student/reader will get a fair idea of on the job operation so that he becomes an asset to the organisation right from the beginning. Thus, the trained worker will save vital resources and the precious time of the organisation which is otherwise *worked on untrained workers* who do not show the expected results *on the staff*. This book will assists guide the reader/student in his daily functioning and thereby enable him to make distinct contribution in the smooth functioning of the organisation where he is employed.

ANIL SAGAR

Contents

Chapter 1

Introduction

The hotel industry has evolved over a long period of time, particularly in a country like India where the guest has always been looked upon as a form of deity who should be respected and extended generous hospitality. That's why in India, at places where even an unknown guest is always welcome, into this day, are called dharamsalas. On the other hand the concept of hotel industry has also established deep roots in our society. And today the hotel industry in India has undergone a sea change and it is contributing to the economy by earning vital foreign currency.

The job of hotel staff demands skilled approach. That's why there is profusion of hotel training institutes. The aim of this book is to guide the candidate who has decided to enter the hotel and catering industry as a professional through the basic stages of food and beverage services and how to develop himself into a skilled worker and an experienced operator. The candidate must be keen, ambitious, interested and have initiative, a pleasant personality and the ability to accept change and should have ability to keep up with modern ideas as they effect his day to day work.

Along with these basic qualities, a thorough and comprehensive grasp of the various aspects of food and beverage service discussed in the book, must enable the candidate to develop into an efficient and result oriented contributor to the development of the organisation for which he works.

Chapter 2

Service - The Product

Hotel is organized and managed by professional who specialise in operations like Accommodation Operations, F&B Service, Production, Engineering, Purchase, Security and so on. As a result, everyone, individually in a group is really accountable for the overall success and quality of the guest experience. The orderlies are accountable for a part of the experience, the Front Office/Reception Staff for another, the F&B staff for another, and so on. In other words, everyone puts his/her best foot forward and produces an experience to the customer so that he gets the value for his money and the organisation gets a repeat customer.

SERVICE AS A PRODUCT

In addition to learning to understand the guest better, we need to understand the concept of service itself better. Though a service is obviously different from a physical product, it is still a product. A Service Product - any incident of *doing for others for a fee* - can be distinguished from a commodity by one or more, and usually several, of the following service characteristics.

1. A service is produced at the instant of delivery; it can't be created in advance or held in readiness.

2. A service cannot be centrally produced, inspected, stockpiled, or warehoused. It is usually delivered wherever the customer is, by staff who are beyond the immediate influence of management.

3. The "Product" cannot be demonstrated, nor can, a sample be sent for customer approval in advance of the service; the provider can show various examples but the customer's own haircut, for example does not yet exist and cannot be shown by the barber, prior to his service.

4. The person receiving the service has nothing tangible; the value of the service depends on his or her personal experience.

5. The experience cannot be sold or passed on to a third party.

6. If improperly performed, a service cannot be "recalled" if it cannot be repeated, then reparations or apologies are the only means of recourse for customer satisfaction.

7. Quality assurance must happen before production of service rather than after production as would be the case in a manufacturing situation.

8. Delivery of the service usually requires human interaction to some degree, guest and staff come into contact in some relatively personal way to create the service.

9. The guest expectations of the service are integral to his or her satisfaction with the outcome. Quality of service is largely a subjective matter.

10. The more people the customer must encounter during the delivery of the service, the less likely it is that he or she will be satisfied with the service.

Note : *We are not suggesting that every service can or should possess every one of these ten characteristics these characteristics paint a picture of a very special kind of transaction between guest and staff (the staff who is working in hotel and service industry) the transaction we call service.*

Chapter 3

Attributes

The waiter who wishes to progress must attain certain standards. It is most important that the service staff creates a good impression on the customer. The following points will assist the service staff in achieving these standards.

Personal hygiene

Since waiting staff deals with food, utmost cleanliness and good grooming is necessary at all times. This applies not only in high class hotels, but in every branch of catering, however humble it may be. Guests are not likely to return to an establishment where staff does not maintain proper hygiene and grooming. Following points should be remembered.

1. Hair should be kept healthy, trimmed and avoid dandruff and ordour. Hair should never fall over the eyes. Waitresses may adopt neat hair styles and particularly ensure a hair length which does not fall on to or below the collar or lengthy hair should be combed and tied neatly and properly.
2. Cleanliness
   (a) Daily bath
   (b) Clothes should be changed daily, especially under garments and socks.
   (c) Frequent washing of clothes is necessary.
3. Care of hands

Trim nails frequently. Always keep nails clean avoid using nail polish.

Use cream to counteract the drying effect of soap and water.

Always wash hand after visiting the toilet.

Sleep

It is better to have regular hours of sleep under good conditions rather than long irregular hours of sleep. Adults require 6-8 hours of sleep daily.

Rooms should be well ventilated and should be free from dust.

Exercise

Exercise is very essential for the normal growth and development of the body and the perfect maintenance of health. It is a means of stimulation of the general physiology of the body, through its effect on the various systems.

Effects of Exercise

1. Exercise gives muscular tone (tone means the ability of the muscle to respond when called upon). Without exercise muscles become flabby thus leading to various diseases. Lack of regular exercise may lead to the weakening of abdominal muscles causing constipation: muscles of trunk and back if not well developed may impair good posture and muscles of back, legs and feet may make walking, running and jumping uneasy and ungraceful.

2. Exercise increases the circulation of blood by its action on heart. This helps in the proper nourishment of the individual cells of the body. Improved circulation of blood also helps in the removal of waste material from the body and regulates body neat.

3. During exercise respiration increases, breathing becomes deeper and the pulmonary circulation is quickened. There is a considerable increase in the amount of oxygen inhaled, CO_2 and water vapour exhaled.

4. Exercise brings about increased accumulation of food and thus creates a demand for food. The appetite is also increased.
5. The mind is refreshed and the power of observation, toleration, etc., develops, fatigue is reduced, followed by relaxation of the body which contribute towards mental freshness.
6. Exercise prevents obesity.

Rules and conditions for Exercise

1. Exercise in fresh air or in well ventilated area.
2. It should be regular and systematic for maximum benefit.
3. Gradually Progress from simple to more complicated movements.
4. Exercise with periods of relaxation to prevent fatigue.
5. It should suit age, sex, occupation and individual needs and capacity.
6. Wear clothing to allow plenty of freedom.
7. Deep breathing and the movement to be unhurried, smooth and rhythmically, with the body in proper posture, correct technique is important.
8. Expert guidance is necessary for correct exercise to balance good and bad effects.

Cleanliness

(a) Skin

A daily bath or shower should be the minimum standard for good waiting staff. Under garments should be changed daily. Talcum powered for body and feet is acceptable, but scent or perfumes must be avoided by all staff because a guest could be allergic to the smell of the perfumes.

Perspiration and excretion of waste matter takes place through innumerable pores in the skin, hence it is most necessary to keep the skin free from the dirt. Soothed sweat glands function property, thus, preventing lungs and kidneys from carrying on extra load.

Perspiration particularly in the folds of the skin if not removed, may cause skin diseases like ringworm, itching, foot rot and impetigo (impetigo is a rapidly spreading pusinfecting of skin which begins as small bristles on the face, scalp, and hands and once begin the infection may spread rapidly by the towel or by finger nails) Prickly heat also results when skin becomes irritated due to excessive perspiration after exposure to sun or heat.

A few drops of disinfectant once in a while helps, those parts which perspire most can be washed in a strong salt solution (3 tea spoon full to one pint of water).

(b) Teeth

Teeth and a clean mouth are vital, both for appearance and a wholesome breath. Teeth should be kept clean by brushing at least twice a day.

(c) Feet

Feet need care, both for comfort and cleanliness. Keep toe nails trim and feet well washed. Corns and other painful blemishes may require treatment by a chiropodist disease. For more severe foot weakness medical advice should be sought. Socks or stockings should be changed and washed daily.

Shoes

Shoes should be well fitting. Waitresses should avoid excessively high and pointed heels and shoes should be daily polished.

Punctuality

Punctuality is very important. If a waiter is continuously late on duty, it shows lack of interest in his work and lack of respect for the management.

The waiter should report on duty before the service is due to commence so that he may check his station, sideboard and have a complete knowledge of the menu. Remember early start is a good start.

Local Knowledge

In the interest of customers the staff should have a certain knowledge of the local area in which they work so that they may be able to advise the guests on the various forms of entertainment offered, the best means of transport to places of interest and so on.

The staff should know about the historical places of the city and how far are the Airport, Bus Terminal/Station, Cinema halls, Commercial Complexes, Railway Stations etc.

Honesty (Integrity)

This is very important for the waiter in his dealings with both the customer and the management. If there is trust and respect in the triangle of waiter/customer/management relationships, then there will be an atmosphere for work which encourages efficiency and a good team spirit amongst the food and service operations. The staff should remember that an increase in sale/ profit will result in an increase in his remunerations.

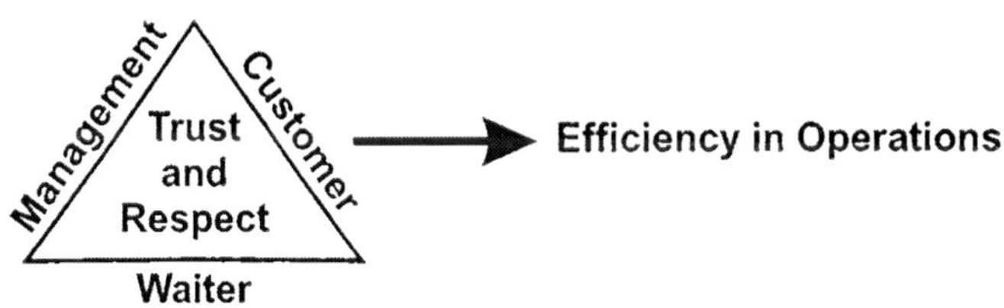

He should take care of all equipment belonging to the establishment and never think that he can deal with it wastefully or carelessly because it is not his own. The staff should remember that an increase in sale/profit will result in an increase in his remuneration.

Courtesy

It is the "Hall Mark" of a good waiter: to be courteous on all occasions. He will certainly be courteous to the customer, but should also carry good manner throughout to the service room and the locker room. His manners should not be just a part of the technique of the restaurant, but inherent in his nature and a sign of well-bred desire to please those with whom he comes into contact.

Every customer, irrespective of his financial standing, should be treated alike. There should never be any frowning on those who may not tip lavishly. Tips should be acknowledged graciously. If placed on the table they should not be removed until the customer has left.

Posture (disposition of parts of body)

Disposition of parts of body is responsible important for the appearance, comfort and efficiency of waiting staff. Standing & walking upright/straight give a good impression to guests and also avoids the body stresses that company slouching.

Uniform

A uniform must always be clean and well pressed. Waitresses blouses and skirts or whatever type of uniform is required, as well as the appropriate aprons, should always be clean and smart. Being clean and smart in appearance gives one the confidence to work well.

Salesmanship

Respectful manners towards customers are highly desirable. A waiter should never be servile for he should be proud of his skill, particularly if he is a good waiter. He is technically a salesman of his establishment and a good salesman should aim to upsell. His voice should be clear, low in pitch and somewhat formed. He should be able to pronounce words properly and express his ideas easily. Since he have to converse with customers, he should acquire a sound knowledge of English because the customers always like to hear a well modulated, pleasing voice with well expressed answers to any questions they may ask.

A knowledge of a second language is of great help, particularly to waiters who seek service in the large hotels or restaurants, where there are often foreign visitors. French is perhaps the most useful for not only are menus written in French, but it is the language used in cookery books and in repertoires of dishes based on *"La cuisine franchase"*.

The waiter works in the front of the house i.e. the food service area and, to large extent, reflects the image of establishment.

He is a salesman and must, therefore, have a complete and sound knowledge of all forms of food and drink and their correct service. Presentation of food and drink is all important and the work of the skilled chef should not be spoilt by the incompetence of the waiters or there may be very quickly a loss of trade and goodwill.

Sense of urgency

Establishment has the maximum amount of business over the service period, with as high a net profit as possible. The waiter must develop a sense of urgency.

Attitude towards guests

The correct approach to the customer is of utmost importance. The waiter must not be servile, but anticipate the guests needs and wishes. A careful watch should be kept on the guests at all the time during the service without staring. Care should always be taken while dealing with the difficult customers. Never argue with the guest which will only aggravate the situation, but refer all complaints to someone in authority in the food service area, i.e. Restaurant Manager or the Head Waiter, if you feel that the situation/problem is out of your reach to amend/handle.

Memory

This is essential to the waiter because it may help him in various ways in his work if he knows the guest or his/her likes and dislikes, for example: Where they may like to sit in the food service area, what are their favourite drinks and the name of guest etc.

Chapter 4

Telephone Manners

Telephone is an instrument for communication. Always remember that on the phone you are speaking to a human being and not to an instrument. Therefore, you are required to generate a certain amount of warmth and you have to judge the mood of the person by his voice, which is, by no means, an easy job. There are certain basic rules which are to be followed while handling phone :

1. When the telephone rings answer it immediately. Answer by saying "Coffee Shop (or your area) Good Morning (Afternoon, Evening). May I assist. You. Do not answer by saying "Hello" or "Ya".

2. If you are attending to a guest when the telephone ring, tell the caller "Coffee Shop" (or your area) Good Morning. Could you please wait a minute," or "I won't keep you waiting for a moment".

   However, if your communication/talk with the person/guest you were speaking when the phone rang seems not to finish soon, give preference to the person waiting on phone because the personally/physically present guest will understand your situation better rather the person on phone.

   You are the best judge to judge the situation and take a decision, which largely depends on your personality. Your objective should be the both person should be attended without making either of them waiting for long. So, be precise - to the point while on duty.

3. Do not leave the caller waiting for too long before attending to him. The caller should never wait longer than 45 seconds without someone getting back to him and saying "I am sorry to keep you waiting".

4. Keep all PERSONAL calls brief, as the telephone is mainly for guest use. They should be discouraged during duty hours unless important.

5. Speak clearly, Pronounce your words carefully because the caller may find it more difficult to understand you on the phone, than when in person.

6. Be polite, friendly, helpful on the telephone, only you voice can identified the volume, we want quest to receive.

7. Be brief and to the point, but not abstract. Remember telephone time costs money and the operator may be holding another line to be connected.

8. Do not eat & chew any thing while on telephone. I make bad impression on the caller.

9. Leave good impression at last also, say "Thank you for calling", before you hang up the phone.

Chapter 5

The Restaurant

Food Service Area

A restaurant is an establishment committed to the sale of Food and beverage. Restaurant may be an independent business entity under individual ownership or a part of a hotel. A Hotel consists of rooms and restaurants for the purpose of rendering services and generating revenue. Restaurant offering a la crate or table d'hote menu using the triplicate menu checking system have a full brigade of waiters to render services to the customer.

Basically restaurants provide tables and chairs for people to rest and sit and eat food prepared by attached kitchen. Restaurant can be defined as "REST-AU-RENT" i.e. a place where a person can rest and pay rent for it.

In order to differentiate between these various forms of restaurants a brief description of each is given below.

Speciality restaurant

Those restaurants which offer speciality cuisines like Chinese, Japanese, Indian, Italian, cuisine etc. would be termed as "Speciality restaurants". The entire atmosphere, concept, theme and type of food served is fully based on cuisine and its original ambience. Generally opens for the 2 major meals speciality restaurants Lunch and Dinner and the staff does a broken shift in these restaurants.

Coffee Shop

Basically a concept from the United States, where a quick service is provided because the guest does not have enough time to wait for the meal. The menu includes various item from light to heavy catering to all categories of customers. Appetizers, soup, main course dishes, Desserts and Beverasges are essentially included in the menu. Here food is served pre-plated and the atmosphere is informal. Table cover layouts are less elaborate and have basic essentials only. It is a place where people either come to have a fast meal or light snacks to compensate for the meals. It is open round the clock.

Night club

Here a very extensive and expensive a la carte menu would be offered by a full brigade of waiters. The night club would be fully licensed, offering a full range of alcoholic beverages. There would be music, dance or cabaret every evening. It is open at night for dinner. It is a place where a person can entertain and eat.

Cafeteria

Here a form of self-service is in operation although there may a section within the cafeteria set aside for waiters service. The number of staff would be limited to those required for service and some staff with trollies for cleaning purposes and counter service hands. The menu would normally be displayed near the entrance and all dishes priced individually. A good presentation and display of food is essential to assist the customer in his choice, and adequate service points to ensuring maximum turnover in a minimum period of time. The cashier would be seated at the end of the service counter to collect payment for the meals served before the customers are seated.

Cafe

Here is usually waiter, service. The menu offered would be a limited table d'hote menu possibly supported by a variety of hot snacks each individually priced. The waiters would make out bill and payment may be made to either the waiters concerned or a cashier. The decor is relatively inexpensive and the service quick.

Snack and milk bars

Here to ensure maximum income and turnover of customers, all equipment required for the service of different dishes and beverages offered would be sited behind the main counter so that it may be operated by the minimum number of staff. This, therefore, cuts down on the labour costs. The customer orders at the counter and receives his order over the counter. He may then seat himself at the table or on a bar stool of the main counter to consume the item ordered. Payment would be made at the conclusion of the meal to the operator behind the main counter or possibly on receipt of the mean over the counter or to waiters generally employed for clearing purpose, relaying and to ensure the payment of all bills. It is very quick and speedy form of service. A wide variety of cold and hot milk drinks would be offered.

Discotheque

A restaurant which is particularly meant for dancing to recorded music. An essential post of a discotheque is a full range of alcoholic beverages, while the food offered consists mainly of snacks.

Buttery

This is sometimes the name give to a restaurant within a luxury hotel. Therefore all those general points about the luxury hotel would apply here. In some instances the a La carte menu would be supported by a small tabled hotel menu and on occasions there are speciality dishes offered from a service point set up within the buttery. This may take the form of a (shell fish 'BAR'). The turnover of customers would be mainly non-resident.

Grill Room

A restaurant which is particularly meant for various meat cuts and grilled or ROAST in front of the guest there is a partition of glass. Between the guest and the kitchen. So that the guest can choose his meat cut and see the actual preparation of food.

Dining Room

This type of restaurant you will found in smaller Hotel or Hospital and may be in motels, usually it is not economical to have more than one eating place. The dining room is basically meant for hospital staff or the residents of the hotel but may be open for out guest also. The kind of service in dining room is buffet service or you can say self service.

Drive In

This type of restaurant you will found on Highways. This type of restaurants have good parking area. They do not have sitting arrangement. You have to take food and sit in car or eat while driving. This is very useful for the service kind of peoples who do have time to sit and eat.

Chapter 6

The Restaurant Staff

F&B
ORGANISATION CHART

Restaurant Managers

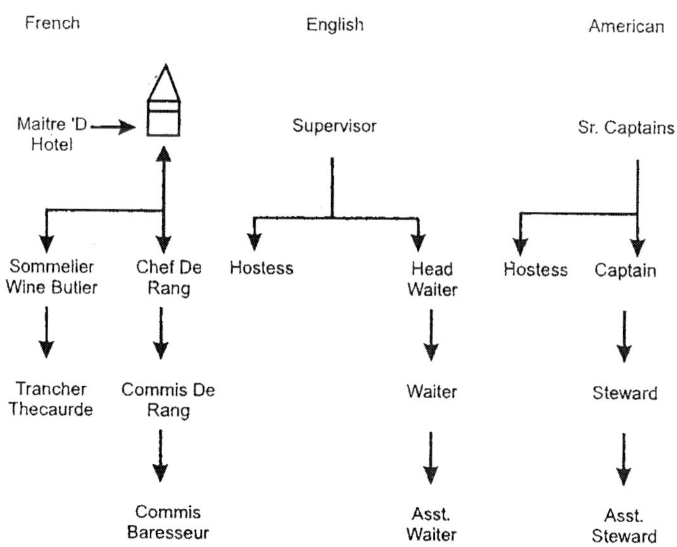

The Restaurant Manager

Has overall responsibility for the organisation and administration of the food and beverage service areas. This includes the lounges, floors, grill rooms, restaurants and possibly some of the private banqueting suites. It is the restaurant manager who sets the standards for service and he is responsible for any staff training that may have to be carried out on or off the job. With the assistance of the head waiter he may make out duty roaster, holiday lists and hours on and off duty so that all the service areas run efficiently and smoothly. All staff would be engaged after interview with the personnel manager and the restaurant manager.

Senior Captain

He has overall charge of the staff team in the restaurant and is responsible for seeing that all the duties necessary for the service are efficiently carried out and nothing is forgotten. He will aid the reception head waiter (Hostess) during the service and possibly take some orders from the guests if station Captain is busy. He helps with the compilation of the duty Rota and holiday lists. He relieves the Restaurant Manager or the Hostess on their day off.

Captain

Captain has the overall responsibility for the team of staff under him and serving a set number of tables, which could be anything from 4-8 in number, from one side board. The set of tables under the control of Captain is called a station. He must have a good knowledge of food and wine and their correct service, and be able to instruct those under him. He will take the order from the guest and carry out all the service at the table with the help of his commi de rang (steward), who is second in command of the station, this includes any gueridon work that may be necessary.

Hostess

Hostess is responsible for accepting any booking and keeping the booking diary up to date. She will reserve table and allocate these reservations to the stations. She greets the guests upon arrival and will escort them to the table and seat them, leaving them in the charge of the station head waiter. The hostess should have good knowledge of food and drink and be able to instruct trainees where necessary. The hostess should have a pleasing personality.

Commis de rang (Steward)

Acts by instruction from the captain. He fetches and carries service of either vegetables, sauces, offering bread rolls, placing plates upon the table and so on. He also help clear the tables after each course. During the pre-preparation period he would carry out some of the cleaning and preparaing tasks.

The Trancheur

The Trancheur is responsible for the carving trolley and the carving of joints at the table as required. He will plate up each portion with the appropriate accompaniment. He has to be very skilled to get the maximum number of portions from each joint with as little wastage as possible and in a very presentable way.

The Sommilier (Wine butler)

Is responsible for the service of all alcoholic drinks during the service of meals. He must have a thorough knowledge of drink he has to serve; of the best wines to got with certain foods and of the licensing laws in respect of the particular establishment and area. He must be a salesman and sell as much liquor as possible as this is one of the main sources of income in an establishment. Wine butler must be a responsible person versed in the skills of shaking and stirring cocktails non-alcoholic drinks and should have, the ingredients necessary for the making of cocktails mocktails ready before the operations.

Tr. Asstt. Steward

Is the learner or has just joined the food service staff and is possibly wishing to take up waiting as a career. During the service he will keep the sideboard well stacked with equipment and he may help to fetch and carry items as required. He would carry out certain cleaning tasks during the pre-preparations periods. The Asstt. Steward may be given the responsibility of service of hors d' oeue, cold sweets or assorted cheese from the appropriate trolleys to the guest.

Chapter 7

Co-ordinating Department of Food and Beverage Service

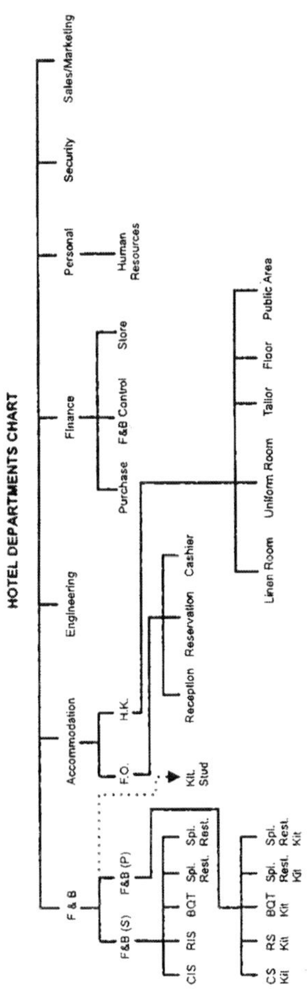

Kitchen

Kitchen is the called as production this is very important for F&B service or you can say production department is the backbone of the Hotel.

Indian Kitchen

This is the place from where you pick up all the Indian food. In Indian kitchen there are two section (i) Grave, and (ii) Tandoor.

(i) *Grave* : All the grave item from the grave section.

(ii) *Tandoor* : All the Tandoori Item or roast and Indian Bread (Roti) from Tandoor only.

Gardé Manger

This is the place in kitchen who should prepare all the cold preparation and salads. This department is responsible for fruit and vegetable carving (to give good shape to the vegetable by knife only e.g., fish, rose, man, any animal) and

F&B Service Hierarchy Chart

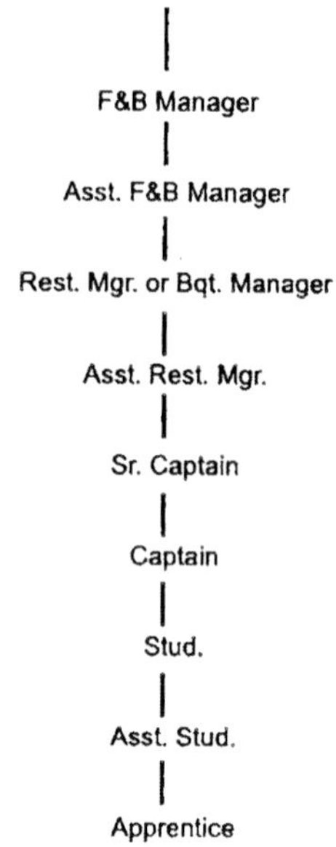

Kitchen Hierarchy Chart

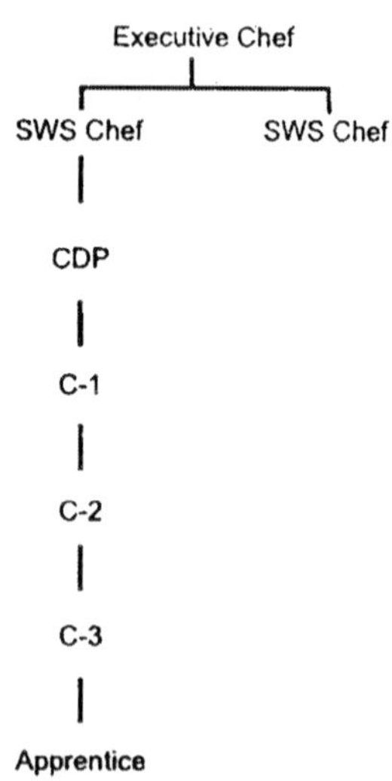

Ice carving which is very popular in banquet's parties in all over world.

Pantry

This is the place from where you can pick up all the fresh juices, sandwich, salad portions individual, in some hotel pantry is giving fruit platters, some hotel giving from Garde Manger only.

Still Room

This is the place for hot beverages. But it depend on the hotel normal hot beverages, from pantry only or you can say the still room is the part of the pantry.

Hot Range & Grills

This is the place for pickup at hotel continental food items such as steak.

Bakery and Confectionary

This is the place to fulfill the requirement of bread, cookies, cakes as per the requirement of the hotel.

Kitchen Stewarding

This is the department who is fully responsible for controls of breakage, storage afall the equipment of F&B and issuing of cutlery and crokery and maintenance of all the service equipment. This department is also responsible for sanitation and hygiene of the Inter hotel it is also responsible and incharge of employees food cafeteria.

Stores

This is the place from F&B pickup all the supplies of proprietary sauces, dry store items, and stationary for restaurant, banquets and F&B.

Liquor Store

This is the place from where F&B pickup all the supplies of liquor and soft drinks. This department is very much under of F&B controls.

F&B Controls

This department is responsible for controls of liquor and soft drinks and unauthorised of food by any staff. This is the department which is also responsible for pel freg and portion controls and the main job of the department is to control food cost.

House Keeping

This department is responsible for the maintenance and the ethnic beauty of the hotel and also responsible for the staff uniform and supplies of linen and flowers to the F&B Department and also responsible for take care of all the plants and agriculture in the hotel. This department is basically known as Hotriculture Department is under House Keeping only. And cleaning, polishing, etc., in the Restaurant and Banquet.

Engineering

This department is responsible for maintenance of the mechanical and electrical equipment in the restaurant.

A/C Department

Cashier : Restaurant cashier receive all the cash and signs bill and money from F&B department.

Front Office

The F/O maintains the records of all the guest staying in the hotel. This department also co-ordinates all group movements and meal plans. If a resident wishes to sign his bill, the steward may contact this department for confirmation of guest's name and room number to fill the correct name and Room no. in Bill.

Chapter 8

Hotel Plans

Different Types of Hotel Plans

There are four major types of plans which are as follows :

1. *European Plan* : This plan includes room rent only and the guest is required to pay the meals consumed.
2. *Continental Plan* : Here the guest pays for the room and the rate includes continental Breakfast.
3. *American Plan* : This plan includes complete boarding and lodging as the room rate also incorporates the charges for all the meals.
4. *Modified American Plan* : As the name suggests, this is a modification of the American Plan. This plan includes the room rent plus continental breakfast plus one major meal.

EP	Room only	Economical Guest
CP	Room + Breakfast	Commercial Guest
AP	Room + Breakfast + Lunch + Dinner	Resorts Tourist
MAP	Room + Breakfast + Lunch or Dinner	Tourist Business Class

Chapter 9

Menu

A Menu is a document prepared by the management for providing information to the customer regarding the variety of food offered and the price being charged for the same. Making a menu is often regarded as setting or solving a cross word puzzle. 'Balance' in gastronomic terms is of course important on 'set' menus for functions, prearranged parties or fixed meals with little or no choice.

In all cases, menu must be realistic in economic terms. In commercial catering they must lead to parties. Whatever the type of food and beverage operation of establishments, menu is a vital piece of business, the menu projects on operations catering intentions.

ORIGIN OF THE MENU

Menu is the French word and english we could 'Bill af fare' in the older time it was not presented at the table instead of menu variety of dishes to be served on the table 8 to 10 dishes before the guest enter in the restaurant. after sitting of guest what guest want he can keep on the table rest all removed from the table. The dishes is consumed by guest it is called "entree" and other removed dishes called "relieves".

In the year of 1541 Duke Henry of Brunwich was did a very good job instead of puting all the dishes on the table he wrote all the dish name on the Big slip and presented to the guest or hang on the wall guest can refer these dishes according his appetite, menu developed from some such an event, the slip of dishes of

Bill of fare was very big and large and it was placed on the wall for everyone to read, as the time change menu became smaller and presentable and in many copies per guest.

POINTS TO CONSIDER WHEN COMPILING A MENU

There are a number of considerations to keep in mind when compiling a menu.

1. *Type*

   (a) Assess the type of meal required.

   (b) Assess the type of kitchen and staff available in relation to equipment and skills.

   (c) Assess the type of the food service area and its number capacity in relation to the china, silver and glassware available, the skills of food service area staff and the number of courses to be served.

2. *Supplies*

   Seasonal and local availability of supplies.

3. *Balance*

   (a) Light to heavy and black to light very the sequence of preparation of each course.

   (b) Change the seasoning flavouring and presentation and ensure that garnishes are as normally with the main dishes.

   (c) Principals of disgestibility.

   (d) Customer's individual performances.

   (e) Season of the year and nature of occasion.

   (f) Resources of the kitchen staff and equipment and what is possible to prepare in time,

   (g) Clear cost and price policy.

   (h) Balancing of foods, not only nutritional and for digestion but also in terms of varied but balanced flavour, colour texture and consistency.

(i) A balanced meal as to colour and ingredients in proper sequence.

(j) For gastronomic reasons, a meal should not contain two dishes which are composed of the same ingredients.

(k) Two white meats and two dark meats should never follow each other. For example, chicken not be followed by fish, or beef by mutton.

(l) A light entree should be followed by a heavier dish.

4. *Food Value*

Use commodities and methods of cooking which will preserve the natural nutritive properties of the raw materials.

5. *Colour*

Avoid either dishes of colour or repetition of similar colour.

6. *Language*

The menu should be written either all in French or English or local language and be easily understood by the customer.

TYPES OF MENU

There are only two basics classes of menu normally i.e., (i) Table d'hote, and (ii) A la carte:

1. *Table d'hote (Table of the host)* : The definition of table d'hote menu is covered by the following points :

(i) The menu has fixed number of courses and there is no choice within each course.

(ii) The price is fixed and the dishes will be provided at a set time.

2. *A La Carte (A dish of the day)* : Some establishments daily carry on a la carte menu called Plat de jour. These dishes are normally priced and potatoes and vegetable are induced in the price. A plate de jour is always a 'ready dish'. It is not a means of getting rid of kitchen left overs.

Specialities however may required pre-ordering. A good waiter should always be aware of the plat de jour and also of reasonable dishes to recommend. Other factors to note in lacking orders and effect helping to sell the food and service of the restaurant includes.

The number of courses on an a la carte menu and dishes within each course depends on the size and class of the establishment. In an establishment where full food preparation and service brigades are in operation a full menu may be offered. In this case the courses or sections at the menu may be divided as follows :

1.	Hors doéuvre	2.	Potages
3.	Oêufs	4.	Farineaux
5.	Poisson	6.	Entrée
7.	Sorbet	8.	Relevé
9.	Rôti	10.	Légumes
11.	Salades	12.	Buffet froid
13.	Entrement	14.	Savoureux
15.	Fromage	16.	Dessert
		17.	Beverages

The number of courses are restricted in modern time to eleven courses. Thus, the French classical menu include.

1.	Hors d oéuvre	2.	Potage
3.	Poisson	4.	Entrée
5.	Relive	6.	Sorbet
7.	Rôti	8.	Legumes
9.	Entremets de sucre	10.	Savoureux
11.	Dessert	12.	Tea & Coffee

FRENCH CLASSICAL COURSE

Today, the normal order in which dishes are served :

1. *Hor's d' oeuvre* : This course is usually composed of tangy salty tit-bits, aimed to stimulate appetite, Horse d' oeuvre varies applies to variety of side dishes offered as appetizers

such as potato, salad, prawns, dives, Russian salad, cold egg dishes. Alternatively an hors d' oeuvre may be a single items served as preliminary appetizer course before the soup for example dishes like melon, caviar, oysters smoked salmon, sausage, salami and smoked ham.

2. *Potage (Soup)* : Two soup are usually served on a table d' hotel dinner, one clear (consomme) and one thick, (creme, veloute or puree) only one soup is served with each meal. The clear soup is listed before thick soup.

3. *Poisson (Fish)* : Two kinds of fish are frequently offered on a table d' hotel dinner. One is invariably a poached fish served with a sauce mousseline or hollandaise or a veloute sauce. Baked/ Fried potatoes are usually served with this type of dish.

4. *Entree* : This is the first of the meat courses. At dinner, it is usually complete in itself in the way that it is accompanied by its own appropriate vegetable and accompaniments garnishes. This course tornadoes and so on. A luncheon entree may be more substantial and additional vegetable may be served separately.

5. *Remove (Releve)* : This could be a large joint such as a saddle of lamb, Cushion of veal, braised ham or even venison. Potatoes and one or two vegetables are served with this course.

6. *Sorbet* : This course is intended to be a pause during a long meal. A sorbet was supposed to 'Settle' dishes already served and stimulate the appetite. It is a water ice, usually flavoured with a sparkling wine and is served in a tall glass with a spoon. Cigarettes, usually Russians are passed at this stage and ten minutes are allowed before the next course, with shorter menus, more disciplined appetite and better culinary knowledge the sorbet is less frequently included and it role is increasingly questioned.

7. *Rôti (Roast)* : This consists of poultry or game, such as chicken, duck turkey, pheasant or partridge served with sauces and gravy. A fine roast such as a tenderloin filled could also be served.

8. *Legume (Vegetable)* : The French customarily serve a finely dressed vegetable as a separate course, for example asparagus served with sauce hollandaise or beurre fonds.

9. *Entrements de sucre (Sweet)* : This consists of a hot and cold sweet such as souffles, Crepes, Coupes.

10. *Savourys (Savoureux)* : A savour usually consists of tit bits on a hot canape of toast or fried Bread. All types of cheese may be offered together with the accompaniments.

11. *Dessert* : All forms of fresh fruits and nuts may be served accompanied by castor sugar and salt. They are sometimes placed on the table as part of *the decorations.*

    Liqueurs offered

    Tea & Coffee

VARIETY OF MEALS

In very recent years 'fast food' has become very popular. This is largely due to the gradual changes in eating habits of people. Of late, people have the tendency to eat when hungry. Some other meals are taken at times and in a style which do not fully accord with traditional ones for example brunch is a meal halfway between breakfast and Luncheon. Some major meal and their French designations are :

1. *Petit Dejeuner (Breakfast)* : This French term includes both English/American and continental breakfasts. In the former case, cooked items are brocaded by fruits, fruit juices or cereals.

2. *Brunch* : A meal between breakfast and lunch, this concept is popular in resorts where people/guest rests are not in any type of hurry.

3. *Lunch (Luncheon)* : Full meal at noon time in the afternoon.

4. *High Tea* : A tea meal, including hot or cold cooked items, between Lunch and Dinner in the afternoon.

5. *Dinner* : Full meal in the evening time.

6. *Supper* : Less formal evening meal taken after dinner or at late night.

Counrt Breakfast

Fruit Juice : Orange, Grapefruit or Tomato

Breakfast Rolls : Croissat ect. or Bread, Tea/Coffee

Cover : Side Plate

Side Knife

Serviette

English Breakfast

Fruit Juice : Orange, Pineapple, Grapefruits

Cereals Breakfast : Porridge with Hot or Cold Milk.

Fish : Grilled Herring, Fried or Grilled Kippers.

Eggs : Fried, Poached, Scrambled, Boiled, Omelette

Meat : Fried or Grilled Bacon, Sausages.

Breads : Toast Rolls, Croissants, Jam, Butter, Marmalade, Tea/Coffe.

APPROXIMATE WAITING TIMES FOR DISHES ON A LA CARTE MENU

These may vary/Differ from organisation to organisation:

Special Hors d'oeure	-	10 mins
Soup	-	3 - 5 mins
Fish (Fired or Grilled)	-	8 - 10 mins
Liver	-	12 - 15 mins
Rump/fillet steaks	-	12 - 15 mins (according to order)
Chateau briand	-	12 - 15 mins
Roast Chicken	-	15 - 18 mins
Lamb cutlets	-	8 - 10 mins

Chops	-	12 - 15 mins
Omelettes	-	4 - 6 mins
Souffles	-	They are ready in bakery
Veal	-	12 - 15 mins

Chapter 10

Type of F&B Organisation

The Hotel Industry provides approximately 40 million meals per day in various type of operation spread over the following groups - viz. Restaurants, snack bar, cafe, Brasseuir, pubs, night clubs, bars, canteen - Institutional & Industrial.

The service of a meal may be carried out in many ways depending on a number of factors, i.e.,

Private caterings Banquestings - Resorts - Hotels - Holiday camps, other tourist or short stay accommodation, pilgrim centres, railway, airway, caterings on ships, take aways, welfare catering etc.

Group

1. The type of catering establishment.
2. The type of customer to be served.
3. The time available for meal.
4. The turnover of customer expected.
5. The type of menu presented.
6. The cost and standard of the meal served.
7. The site of the establishment.

Group	*Origin*
Hotels, Restaurants and other tourist accommodation	Developed from Inns & Dharamshalas supported by wheel & more people travelling more often.
Fast food, take aways, Retent shops/super markets	Influenced by Americans, supported by industrial development resulting in people having less time at disposal for meals.
Banqueting/Conferences Exhibition (Caterings - Industrial & Institutional)	Initially part of Hotel but the sheer magnitude of business potential have forced it to become a major group rightfully/in its own right.
Transport Catering Railways/ Airlines/Marine/Roadways	Developed because people need to eat while travelling, supported by the circumstances providing after Industrial revolution when the masses started travelling.
Catering Industrial/ Institutional (School, College, University)/Welfare (Prison/ Pilgrim centre/places)	Influenced by the belief that better fed workers work better
Outdoor Catering Liesure attraction	Developed because of need to provide special services at special occasions.

The service can be better explained if we consider if from the customer point of view or the work flow of business activity at a F&B outlet. The customer enters the F&B area — Order the food — Service taken place — consumption payment.

The consumption can take place before payment or after payment depending on the type of establishment and service provided.

Chapter 11

Type of Service

French Service

It is very personalized service. Food is brought from the kitchen in dishes and salvers which are placed directly on the table. The plates are kept heated, the dish and the guests help themselves.

English Service

Often referred to as the 'Host Service' because the host plays an active role in the service. Food is brought on platter by the waiter and shown to the host for approval. The waiter places the platters on the table. The host either portions the food into the guest plates directly or portions the food and allow the waiter to serve.

Silver service

The table is set for horsd ocuvres, soup, main courses and sweet dish in sterling silver ware. The food is portioned into silver platter at kitchen itself which are placed at the sideboard with burners of hot plates to keep the food warm in the restaurant. Plates are placed before the guests. The waiter then picks the platters from the Hot plate and present the dish to the host for approval spoon and fork. All food is presented in silver dishes with elaborate dressing.

American Service

The American service is a pre-plated service which means that the food is served into the guests plate in the kitchen itself and brought to the guest. The portion is pre-determined by the kitchen and the accompaniment served with the dish balance the entire in terms of nutrition and colour. This type of service is commonly used in a coffee shop where service is required to be fast.

Grill Room Service

In this form of service various meats are grilled in front of the guest. The meats may be displayed behind a glass partition counter so that the guest can select his exact cut of meats. The food comes pre-plated.

Buffet Service

Self service where food is displayed on tables. The guest takes plate from a stack at the end of each table of requests the waiter behind the buffet table to serve him for sit down buffet service. Tables are laid with crockery and cutlery as in a restaurant. The guest may serve himself at the buffet table and return to eat at the guest table laid out. A few courses like the appetiser and soup may be served at the table by the waiter.

Russian service

An elaborate silver service much on the lines of French service except that the food is portioned and curved by the waiter at the gueridon trolley in the restaurant in full view of the guests. Display and presentation are a major part of this service. The principal involved is to have whole jaunts. Poultry grave and fish elaborately dressed and garnished presented to guests and carved and portioned by the waiter.

Gueridon Service

This is a service where a dish comes partially prepared from the kitchen to be completed in the restaurant by the waiter. The cooking is done on a gueridon trolley which is a mobile trolley with a gas cylinder and burners. The waiter plays a prominent part as he is required to fillet carver, flambe and prepare the food with showmanship. The waiter has to have considerable dexterity and skills.

Service Equipment

The general points which must be considered when purchasing equipment food and beverage service area as follows:

1. Flexibility of use
2. Design
3. Durability
4. Ease of maintenance
5. Storage
6. Colour
7. Type of customer
8. Type of service to be offered
9. Stackability
10. Cost, funds available
11. Availability in the future
12. Rate of breakage (China)
13. The site or location

Chapter 12

Linen

It depends on the organisation, management, type of restaurant, etc. Following are some examples commonly used in the industry but they may not be necessarily a standard for a hotel restaurant.

1. Table cloths

(a)	2.6 feet square table	54″ × 54″
(b)	3 feet 137 cm × 137 cm square table	72″ × 72″
(c)	183 cm × 244 cm rectangular shaped tables	72″ × 96″
(d)	183 cm × 137 cm rectangular shaped tables	72″ × 54″

2. Slipcloth

1 m × 1 m (to cover a "grubby" tablecloth) 36″ × 36″
To cover a "grubby" table cloth

3. Servilities Napkins

Should be square 21 m × 21 m

4. Buffetcloths

Minimum size 2 m × 4 m = 6 feet × 12 feet

5. Waiter cloths or service cloths

Used by every waiter as protection against heat and to keep uniforms clean.

6. Tea and glass cloths

The best are made of Linen or cotton.

Chapter 13

Furniture

Following are some standard sizes of the restaurant furniture. But every restaurant and hotel is not bound to follow them. Each have their own concept and it may differ according to the decore and theme of the particular restaurant.

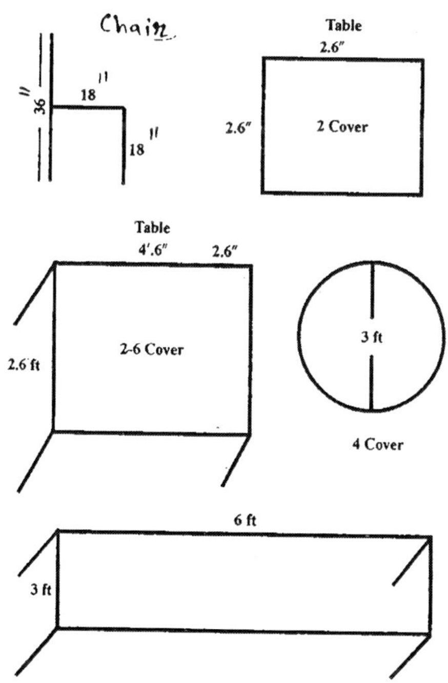

Figure

Chapter 14

The Sideboard/ Demi Waiter/ Dummy waiter

The style and design of a sideboard varies from establishment to establishment. It depends upon :

1. The style of service and the menu offered.
2. The number of waiters working from one sideboard.
3. The number of the tables to be served from one sideboard.
4. The amount of equipment it is expected to hold.

It is essential that the sideboard is of minimum size and portable so that it may be moved easily if necessary. If the sideboard is too large for its purpose, it is then taking up space which could be used to seat more customers. The sideboard is used to hold all the equipment required by a waiter during the service of a meal. The design of sideboard varies from one establishment to another. The welled compartment holds a hot plate. Ensuring it will be in the same level as the rest of the working surface is an important safely consideration. The cutlery drawers should hold, from right to left.

The top should be of a neat resistant material which can be easily washed down. If a hot plate is to be used then it should be inserted in the top, so that it is in the level of the top and blends with the working job.

After service the sideboard is either completely emptied out or restocked for the next service. In some establishments the waiters are responsible for their own equipment on their station. After service they restorck their sideboard and it is then locked. Where this system is carried out the sideboard also carries its own stock, of linen. In other words, everything necessary to equip a particular waiter, station or set of table is stacked in the Sideboards.

This is only a suggestion order :

- Service spoons and forks
- Dessert spoons, forks
- Soup, tea and coffee spoons
- Fish knives and forks
- Meat knives
- Side knives

The lower compartment is used for dirty lines; the hole in the back of the sideboard above the welled hot plate area is for an electric telex to be passed through; and on the right-hand side is a tray rack. The sideboard may have fitted castors for easy moving.

Diagram of a Side Board

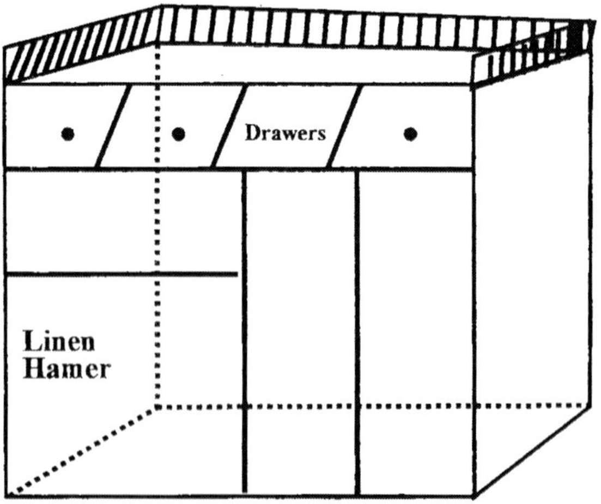

Figure

Chapter 15

Chinaware

BONE CHINA

This is very fine and hard china and is very expensive. The decorations to be found under the glaze only. It can if necessary be made to thicker specifications, if requested for hotel use. The price of bone China puts it out of reach of the majority of everyday charterers, and only a few of the top class hotels and restaurants would use it. Obviously price of a meal in these establishments is high in order to cover this aspect of the overhead. The range of design, pattern and colour is very wide and there is something to suit all occasions and situations.

EARTHEN WARE

Earthen ware is particularly economical where it is in continuous use 24 hours a day, has heavy handling with a high turnover and where hygiene is particularly important.

China should be stored on shelves in pales of approximately two dozen. Any higher may result in their toppling over. They should be stored at a convenient height for placing on and removing from the shelves without any fear of accidents occurring. If possible, keep covered to prevent dust and germs settling on the China.

It is made of Silica, China clay, Soda ash, glazed to give a fine finish. Chinaware is more resistant to heat than glassware.

There are some standard sizes of Chinaware which we use in hotel.

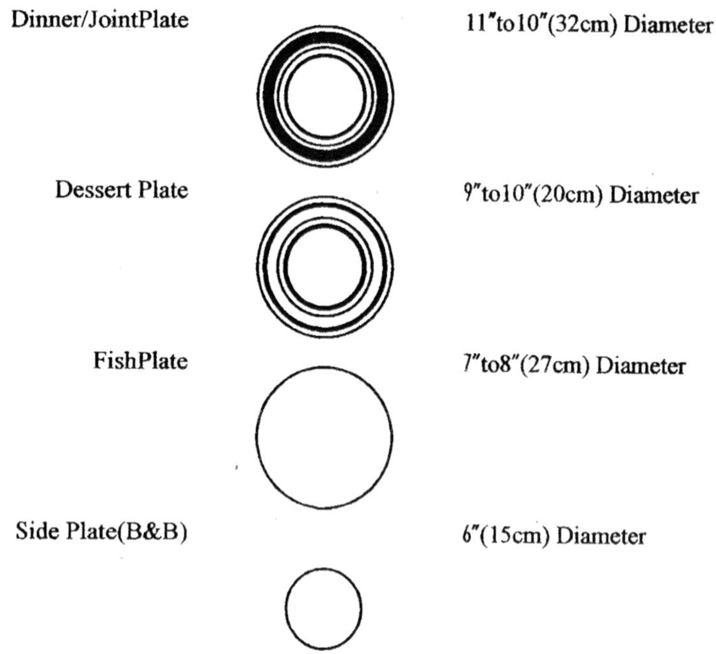
Dinner/JointPlate
11"to10"(32cm) Diameter
Dessert Plate
9"to10"(20cm) Diameter
FishPlate
7"to8"(27cm) Diameter
Side Plate(B&B)
6"(15cm) Diameter

Figure

Coffee cup 8-10 OZ volume

Tea cup 6 1/2 OZ volume

Coffee Demi-Tasse 3 1/2 OZ

Figure

OTHER CHINAWARE

Hot water jug	Butter dishes
Milk jugs	Ashtrays
Cream jugs	Egg cup
Coffee pots	Soup bowl
Hot milk jugs	Platter
Sugar basin	Tea pot
Slop basin	Creamer
Flower Vases	

GLASS WARE (MAGNESIA AND ALUMINA)

The raw materials used are silica and soda ash lead is added to make the glass. When purchasing glassware it should be ensured that it is completely transparent, free of air bubbles and not chipped. Glasses are measured in terms of capacity, anuses or centilitres. Restaurant glassware is usually plain except in speciality restaurant where they may be coloured.

Tips on Glassware

1. Glasses should be stored inverted in single rows with a paper on shelf to avoid slippage.
2. Racks with individual compartments for each glass is a better method for storage (as they can be stacked) and transported.
3. These racks are plastic or rubber lined.
4. In a restaurant, glass must be transported on trays with a tray cloth to avoid slippage.
5. Glasses with a stem must be inverted and held by the stem as shown.
6. Service glasses must be cleaned and held against a light before use for traces of smudges, etc.

CRYSTAL GLASSWARE

Consists of a large amount of lead oxide alongwith Potassium oxide and silica. This is very expensive glassware and should be handled with care.

	Glasses	*Qty.*	*Use*
1.	Water Goblet	10 oz	Drinking water
2.	Red Wine Glass	7 oz	Red Wine
3.	Champagne Tulip	9 oz	Champagne during formal occasions and with food
4.	Parfait Glass	5 oz	Whisky sour, Gin sour, Black Russian
5.	White Wine Glass	5½ oz	White wine and Rose
6.	Wine Goblet	6 oz	Any wine can serve in it.
7.	Brandy Inhaler	9 oz	Cognoc and Brandy
8.	Beer Goblet	14 oz	Beer
9.	Sherry	4 oz	Sherry and Port vermouth
10.	Beer Mug	14 oz	Beer
11.	Champagne Saucer	6 oz	Champagne for a celebration, Daiquiri, Pink lady, Grass hopper, Side car-white lady
12.	Cocktail glass	4 oz	Cocktails
13.	Rolly Polly	9 oz	Drinks on the rocks rusty nails salty dog
14.	Liqueur cordial	3 oz	For liqueur
15.	Old fashion	0 oz	Bloody marry
16.	Pool glass	10 oz	Anything on swimming pool. It should be unbreakable.
17.	Tom Collins glass	12 oz	Fresh lime soda, fruit punch, lassi
18.	High ball glass	8½ oz	Water and spirit example, whisky, gin, vodka, rum
19.	Pony tumbler/ juice glass	5 oz	Juice is breakfast in Banquet.

After using liquids such as tea,coffee,milke etc. holloware should be emptied at once.

Figure

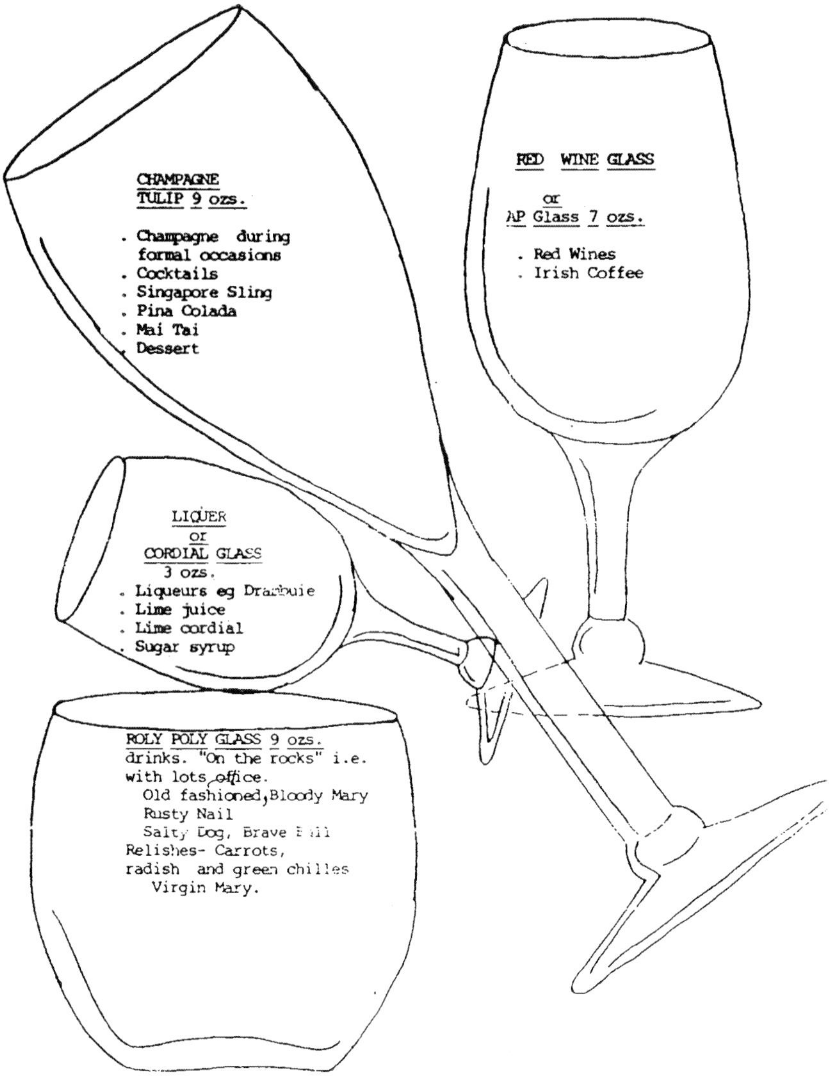
CHAMPAGNE
TULIP 9 ozs.
. Champagne during formal occasions
. Cocktails
. Singapore Sling
. Pina Colada
. Mai Tai
. Dessert
RED WINE GLASS
or
AP Glass 7 ozs.
. Red Wines
. Irish Coffee
LIQUER
or
CORDIAL GLASS
3 ozs.
. Liqueurs eg Drambuie
. Lime juice
. Lime cordial
. Sugar syrup
ROLY POLY GLASS 9 ozs.
drinks. "On the rocks" i.e.
with lots of ice.
Old fashioned, Bloody Mary
Rusty Nail
Salty Dog, Brave
Relishes- Carrots,
radish and green chilles
Virgin Mary.

Figure

WHITE WINE GLASS
5 1/2 ozs.
OR
HOCK
(Hock is the English name
for German White Wines
from the Rhine)
. White Wines
. Rose Wines
SHERRY GLASS
4 ozs.
.Sherry eg. Bristol cream
.Port eg Taylors
.Aperitiffs like Vermouth
.Cinzano Bianco (white)
.Cinzano Rosso (red)
.Martini Bianco
& Rosso
Moilly
Prat
Dubo-
nnet.
COCKTAIL GLASS
4 ozs.
. Cocktails
Dry Martini,Gimlet
Rob Roy,Gibson
Gin Sling
Frappe (liquour in
crushed ice)
COLLINS
12 ozs.
.Juices served 'King size'
.Fresh lime soda/water
sweet/salted/plain
.Aerated waters with lots
of ice
.Lassi - salted/sweet/
plain
.Beer (Room Service)
.Milk Shake
.Squashes
.Fruit punch
.Ice Cream, Soda
.Cold Coffee
.Cocktails
.Tom Collins.

Figure

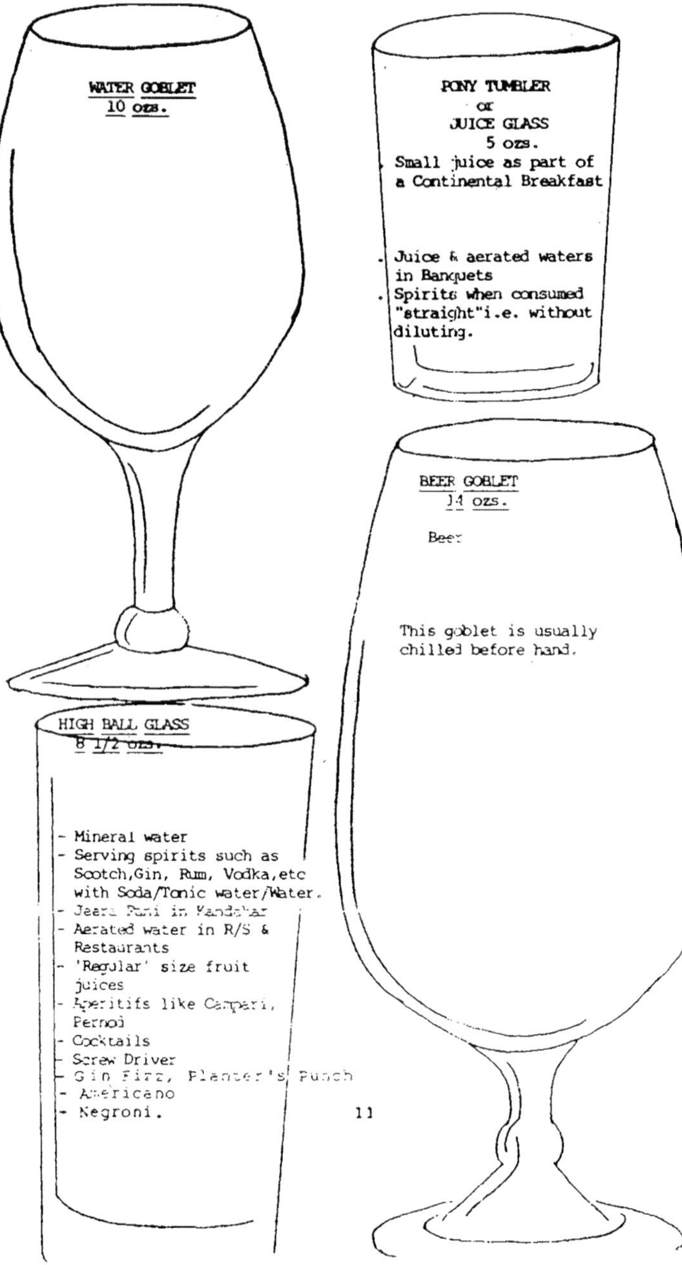
WATER GOBLET
10 ozs.
PONY TUMBLER
or
JUICE GLASS
5 ozs.
. Small juice as part of a Continental Breakfast
. Juice & aerated waters in Banquets
. Spirits when consumed "straight"i.e. without diluting.
BEER GOBLET
14 ozs.
Beer
This goblet is usually chilled before hand.
HIGH BALL GLASS
8 1/2 ozs.
- Mineral water
- Serving spirits such as Scotch,Gin, Rum, Vodka,etc with Soda/Tonic water/Water.
- Jeera Pani in Kandahar
- Aerated water in R/S & Restaurants
- 'Regular' size fruit juices
- Aperitifs like Campari, Pernod
- Cocktails
- Screw Driver
- Gin Fizz, Planter's Punch
- Americano
- Negroni.
11

Figure

Chapter 16

Death in the Hotel (Resident)

1. Call the Hotel Doctor - for medical opinion and certification about the death and possible hour of death in case of a Heart failure, etc.

2. If patient is still alive - call the Hotel Doctor, in the meantime try and provide medical help on the advise of the Doctor or if desired by the patient/guest - remove him to the nearest possible hospital.

   (i) Inform General Manager.

   (ii) Inform Police (advice of Hotel Doctor/ General Manager is absolutely necessary).

   (iii) Inform Embassy/ High Commission/ Consulate, etc., in case of a Foreigner - for the disposal of dead body inform the next of kin.

   (iv) In case of an Indian Guest - inform local contact if any or inform next of kin as soon as possible.

   (v) Dead body should be sent through the Service Elevator and back of the House.

Chapter 17

Fire in Hotel

1. Telephone Exchange gets the first information on the basis of a fire alarm.
2. Information is passed on to L/Manager Immediately.
3. General Manager to be informed at once.
4. Fire Officer of the Hotel to be informed.
5. Chief Security Officer to be informed.
6. Fire Brigade to be informed - only as per instructions of the General Manager, Fire Officer, etc.
7. Evacuation of the Guests through staircases only. Elevators not to be used - due to any possible electrical failure or excessive smoke causing suffocation.
8. Casualties if any - to be referred to the Hospitals.
9. Dead bodies to be disposed of as per instruction of General Manager.
10. Extent of damage to the Guest property and Hotel property to be assessed.
11. In case of extensive damage to the Hotel Property/Guest Rooms - other Hotels to be contacted for alternative accommodation for the Hotel Guests.

Chapter 18

Table Ware (Flat Ware, Cutlery and Hollow Ware)

Flat Ware

In the catering trade denotes all forms of spoons and forks.

Cutlery

In this refers to knives and other cutting implements.

Hollow Ware

Consists of any items made from silver apart from flat ware and cutlery. Tea pots, milk jugs, sugar basins, oval flats.

Asparagus tongs	Nut crackers
Gateaux slice	Grape scissors
Pastry knife/fork	Lemon press
Con-on-con boulders	Snail dish
Lobster pick	Snail fork
Butter knife	Cheese knife
Grapefruit spoon	Skewer (Kababs)
Ice cream spoon	Caviare knife
Sundae spoon	Dessert knife
Snail tongs	Dessert fork

	Equipment	*Dish*
1.	Oyster fork	Shellfish cocktail/oysters
2.	Grape fruit spoon	Grapefruit halves
3.	Steak knife	Grilled meats
4.	Corn-on-the-cob holders	Con-on-the-cob one to price each of the cob
5.	Cheese Knife	Cheese board
6.	Fruit knife and fork	Dessert-fruit basket
7.	Pastry slice	Sweet trolley-serving portions of gasteaur
8.	Grape scissors	To cut and hold a portion of graps
9.	Nut crackers	Dessert-fruit basket
10.	Jam spoon	Preserves
11.	Sundae spoon	Ice-cream sweet in a tall glass
12.	Lobster crackers	To break the shell to gain access to meat
13.	Lobsters pick	To extract the flesh from the claw
14.	Finger bowl	Part of the make up of a cover to certain dishes, oysters, dessert, globe artichoke
15.	Carviar knife	Has a short board blade used for spreading the caviar.
16.	Snail tongs	To hold the snail shell.
17.	Snail dish	Round with two ears. Has 6 indentations to hold a portion (6) of snails.
18.	Snail Fork	To extract the snail from its shell
19.	Silver Skewers	For attractive presentation of Kabab
20.	Pastry knife and fork	Used in service of afternoon teas.
21.	Shellfish cocktail holder	In which a shellfish cocktail would be presented

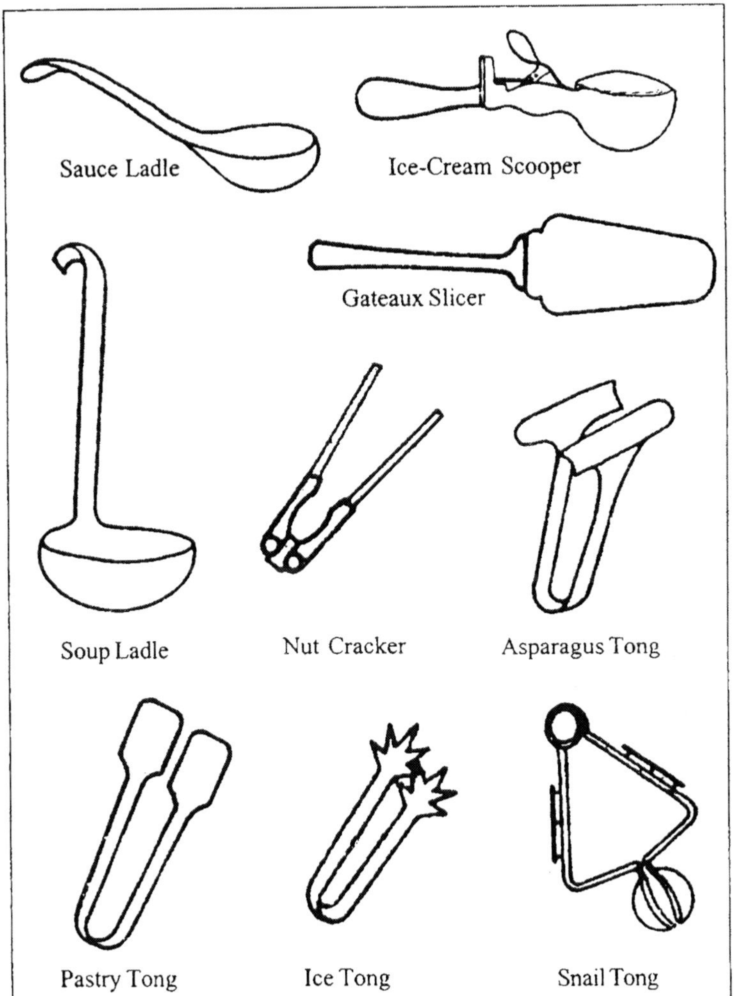
Sauce Ladle
Ice-Cream Scooper
Gateaux Slicer
Soup Ladle
Nut Cracker
Asparagus Tong
Pastry Tong
Ice Tong
Snail Tong

Figure

Chapter 19

The Cover

One of the technical terms very often used in the catering industry is a cover. What does this mean? There are two definitions according to context. Each cover requires 24″ × 18″ of space. Cover is the space required on the table for one person is the beginning of the meal for keeping cutlery, glasses, plates and linen.

"Cover is the minimum area required by a person to sit and eat comfortably."

1. An a lar carte cover.
2. A table d'hote cover.

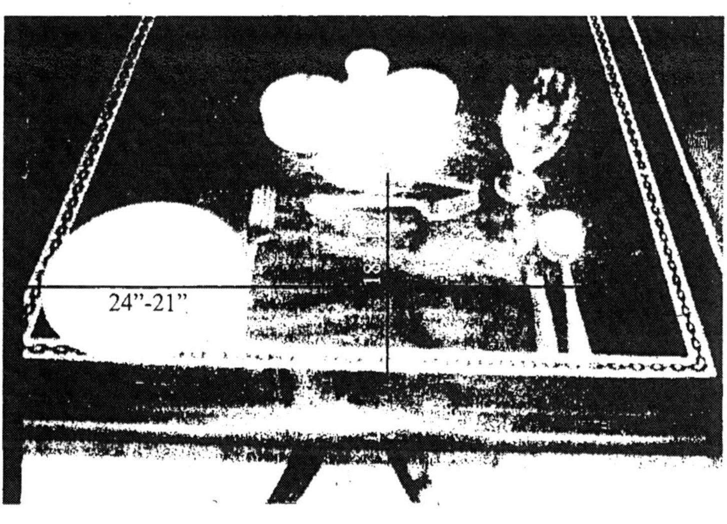

1. An a lar Carte cover

This is the place-setting normally laid in a good class restaurant dining room or hotel in readings for service of either luncheon or dinner.

It would consist of the following :

1. Fish plate	5. Side plate
2. Serviette	6. Side knife
3. Fish Knife	7. Wine glass
4. Fish fork	8. Water goblet

2. A table d'hote cover

This is a place-setting normally laid in a good class restaurant dining room, hotel, medium sized hotel, etc. in readiness for service of either Luncheon of Dinner. It would consist of the following :

1. Serviette	7. Dessert spoon
2. Soup spoon	8. Dessort fork
3. Fish fork	9. Side plate
4. Fish knife	10. Side knife
5. Joint knife	11. Wine glass
6. Joint fork	12. Water goblet

Table Accompaniments

The table accompaniments require to complete the table lay-up are the same whether an a la carte or table d'hote cover has been laid. These would be as follows :

1. Cruet: Salt, Pepper, Mustard and mustard spoon
2. Ashtray
3. Table number
4. Vase of flowers

Characteristics of a cover

1. Each over must be well balanced and should never be crowded with cutlery, glasses and china.
2. All cutlery and other table accompaniments must be placed at least 1/2" away from the edge of the table.
3. Knives and spoons are placed towards the right of the plate and all forks on the left. If no knife is to be used then the fork is to be placed on the right.
4. See that the cutting edge of the knives is always toward the plate with the exception of side knife, the cutting edge of which is away from the plate.
5. The water, tumbler must be on the tip of large knife.
6. The butter dish must be on the top of the forks, alongwith the butter knife of an under plate.
7. The serviette must either be placed in the centre of the cover or on the side plates.
8. Cruet seat must be placed on top of the cover.

After the tables have been laid, another equally important feature that needs careful attention is the completion of the side-board an ideal side-board contain :

1. Cold water in a jug with an under plate and covered with a napkin,
2. All important proprietary sauces,
3. Tooth picks in a Tooth Pick holder,
4. Sugar bowls-one each for white sugar and brown sugar complete with tea spoon,
5. Sugar dredger,
6. Hot plate,
7. Straw holder with straws,
8. Pickles and chutneys,

9. Bread loaf with assortment of rools and sticks,
10. Ashtray-cleaned and polished,
11. Service spoons and service forks,
12. A few pieces of each type at cutlery seed on table,
13. Salvers,
14. Under flats,
15. Half plates, quarter plates, large plates to be used as under plates or as cold plates,
16. Sauces of all types,
17. A crumbling plate (a large plate),
18. A platter - a large plate covered with a napkin, and
19. Linen-spare linen of all types.

Chapter 20

Issuing of Equipment

SILVERWARE

Cutlery and special Hollow Ware items

(Water Jugs, Sugar Bowls, Ash Trays, Bud vases, Candle Stand, Tea Straines, etc.)

Operational par stock of these items to be worked out in coordination with the outlet Manager and approved by the Food and Beverage Manager. These stocks are then issued (against a proper requisition duly authorized) directly to the particular outlet. The outlet Manager is solely responsible for such items as these items are always stored in the premises of Restaurant/ Room service area.

Washing of these item is the responsibility of Kitchen Stewarding but the wiping and bussing to the inside of Restaurants to be done by Restaurant Staff.

Daily polishing of Cutlery is to be done by the Restaurant Staff whereas polishing of Hollow Ware will be undertaken by Kitchen Stewarding Staff but the service staff is to handover the Hollow-ware under direct charge to the Kitchen Stewarding Staff on the specified time and take it back after it has been polished.

General Hollow Ware Items

(Curry Bowls, Platters, etc.)

These items are allocated and issued (quantity worked out in co-ordination with Outlet Manager, Chef and approved by F&B

Manager). To the restaurant and are under the joint custody of Kitchen Staff and Outlet. These items are kept at the pick up points in the Kitchen for easy accessibility of production and service staff.

Washing, wiping, bussing of these items to various pick up points and the upkeep is the sole responsibility of Kitchen Stewarding Department.

China Ware

Operational Par Stocks to be worked out in co-ordination with the outlet Manager and approved by F&B Manager and issued to the particular Outlet.

Procedure for Coffee Shop and Room Service

China Ware : Dinner Plates and Dessert Plates - to be washed at the Dishwashing Machine, wiped and bussed to the pick up counters by Kitchen Stewarding Staff.

B&B Plates, Tea/ Coffee Cups, Saucers, etc., to be washed at the Dishwashing Machines by Kitchen Stewarding Staff. Wiping and bussing of this is the responsibility of the particular Outlet Staff, as these items are stores in the Outlet premises only.

INVENTORY CONTROL

Purpose

It serves as a guide line for the outline Manager and Manager Kitchen Stewarding to avoid serious breakdown in operation. Help in estimating operational requirements.

Procedure

All the Restaurants are required to maintain a Inventory Register. Inventory of silverware, glassware, chinaware and other special equipment which is under their custody.

The Asstt. Manager or Sr. Captain is made responsible for the particular outlet's inventory. He ensures that all the equipment belonging to the outlet is displayed at the time of visit of inventory team, comprising of accounts and Kitchen Stewarding Personnel.

RECORDING AND CHARGING OF SPOILAGE AND BREAKAGE

Purpose

To keep a control over spoilage and breakage and minimise the expenses.

Procedure

Any spoilage or breakage must be recorded and approved. The matter must be investigated and if the breakage has resulted due to negligence or carelessness of the person concerned must be charged and matter should be reported to the Food and Beverage Manager.

Chapter 21

Techniques of Service

Techniques and waiting procedures over the years have emerged for convenience, both for customers need and in aiding waiters to operate productivity and profitably and to merchandise meals. These techniques and procedures should not be regarded as a 'haly writ'. Technical slips will either be readily overlooked if a waiter's contact in other respects is agreeably up to the mark. In applying techniques a waiter must not forget to be courteous and tactful with difficult customers and above all; he must never exhibit impatience and short temper.

Waiter should never amply denigrate the quality of food he is having if he believes dishes are not satisfactory. He should refer to his senior captain or other supervisor before serving it so that a decision may be made about its return to the kitchen. It is important to establish good communication and relationship with guests, notify them of unavoidable delay, explain why, and avoid blaming others.

General practice during service

Above all, the service staff must remember to keep the station tidy. The important factors are DO NOT allow dishes to accumulate and NEVER put anything on the floor.

Use of sideboard

The design and size of sideboard may vary from one organisation to another but the purpose universally remains the same. One must change the styles of service according to the size.

Side stations in out hotel are quite spacious and provide enough storage for silverware, glassware, linen and other accessories such as a sauces, toothpick holders, straw, menu cards, etc. During the service, items brought from the kitchen are placed on the sideboard and after picking up additional crockery, silver, etc. are then served to the guest. The sideboard also has a soil linen storage rack which is used extensively during operation. The linen is later changed by the steward responsible for the job.

Dropping things on the floor

Pick up immediately any article dropped on the floor for return to wash up. They must not be used at the table. Never put any food dropped on the floor on the dish with the rest of the food. Pick it up smoothly and take it right out of the restaurant. Do this with due regard to cleanliness and delicacy. For example, pick up the food with a discarded menu, an aspic or scoop so as to make it clear that the dropped food will be thrown away as guest is likely to watch to see what you are going to do with it.

Cracked and chipped items

Never lay a chipped glass or plate, and avoid serving from any such imperfect article. Take anything dropped on the floor to the side table for replacement and then take it to the wash up.

Items not to be carried

Articles such as teaspoons, etc. should never be carried in the pocket. These should always be stored in a suitable compartments on the side stations. Never carried K.O.T. in pocket.

Handling

Handle all equipment as quickly and without noise. Pick up all equipment by their handles, never by the eating or drinking fast. When serving a beverage keep the jug or pot handle away from the customer but the handle of the cup, etc. from which he will drink should be turned to the customer's right side. Place items such as sugar, cream, milk, etc. to the right of the customer within easy reach.

Sides for services

In many smaller establishment no matter which side of guest should a waiter serve from is given lesser importance. However, in large hotels this has been outlined in principle. Once it was widely accepted that food service should be from the left and the food clearance from the right. Although this is the logically correct practice, yet it might be true to say that clearing from the right is. hazardous to glasses or the table and awkward when guests are drinking. Hence, procedure tend to very according to different establishment's ideas of efficiency and convenience.

However, there are two broad classifications, such as the continental tradition and the English adaptation.

In Continental the main rules are :

1. Place clean plates and glasses from the guests right,
2. Place coffee cups and saucers from the guest's right.
3. Service food from the guests right.
4. Served drinks including wine and coffee from the guests right.
5. Clear all used items such as plates, cups, glasses from the right.

English Adaptation

The main rules are :

1. Plates or utensils for food are placed from the left.
2. Food is served from the left.
3. Used food plates are cleared from the left.
4. Glasses are placed from the right.
5. Orients are served from the right.
6. Used glasses are cleared from the right.

In India, the continental tradition is more popularly followed as it is most suitable.

Waiter's service cloth

The waiter's service cloth should be used only for the purpose intended. The service cloth should always be carried by the waiters during service as the use of this is absolutely necessary while presenting a dish or placing plates on the tables. The use also includes polishing glasses cutlery plates, etc.

While using the waiters cloth for wiping, ensure that the fingers do not touch the glass (or china). Polish both sides of the plate. Wipe plates as they are taken from the plate.

Use of Service cloth

1. Wiping spillage from a place or cup room.
2. Wiping the underside of a dish before placing it on the table.
3. Handling dishes hot or cold.
4. Carrying a tray, using the cloth as a pad as this protects the hand. When in use, carry the service cloth neatly folded on left arm.

Note : *Never misuse the service cloth for wiping furniture, tables to clean shoes, or to flick crumbs of the table.*

Do not allow service cloths to get too dirty. If cloth gets dirty or soiled during the service, discard it at once and take a clean one from the service station.

Serving spoon and fork

One handed manipulation of service spoon and fork should not be ordered as a ritual. Some items, such as potatoes, peas or some other vegetables, are served better using only the service spoon. Fish, fillets, etc., are, however, best served with splayed forks.

Not suing a serving spoon and fork

1. Take a clean set of spoon and fork for each dish to be served.
2. Do not use them again to serve another table.
3. Do not use fork individually, except in case of smoked salmon.

4. Never use two spoon.
5. Never use the same spoon and fork for both potatoes and green vegetable.
6. Do not lift food, do so carefully, so as to ensure that food does not slip or gall on the guests plate.

Serving from Salver

The term salver also includes high quality steel equipment. Similar techniques also apply while serving from copper or other ware.

Prior to picking food from the kitchen ensure:

1. It is the right order.
2. Portions are adequate.
3. The dish is neat and clean.

If any customer have to be served with dishes and others cold dishes serve the cold dishes first.

Presenting a dish

Present each dish at the table prior to serving it. First show to the host for approval. When presenting a dish, lift its over to avoid condensation dripping on to the carpet or table cloth. Dishes are always presented on an under dish or liner.

An oval dish lengthwise on the palm of the hand. Do not serve any dishes hot or cold wihout a service cloth.

Serving from a dish

With food already portioned for service on a silver dish place on with a service spoon and fork stand on the guest's left, feet together with the dish on the palm of the left hand (protected by the pad of folded waiter cloth) : Advance you left food forward, and bring the dish quite close to the level of the guest's plate just over the rim. Take the spoon and fork in your right hand to serve food on the plate.

Serving from a large flat salver

When serving from a large flat dish always remember to :

1. Ensure that guests plates are hot and clean;
2. Protect yourself and plates with a service cloth when carrying a pile of hot plate;
3. Stand on the guest left; and
4. Serve with care to avoid spillage gravy or sauce on the plates rim.

Food on the guest plate

Arrange food on the guest plate appetizingly. Fish or meat is served on the lower centre with vegetable on one side and potatoes on the other. Separate sauces and accompaniments at the top right hand (2 O'clock).

When serving sauce, always check with the guest before pouring it over the main meat. Arrangement of food on the guests plates should be aimed towards reproducing the effect achieved by the chef on the salver dish. The chef has to make the dish attractive. And food not arrange properly result in an uninteresting and unappetizing food.

Lamps and hot plates

Side boards are often equipped with hot plates. These may be necessary to keep dishes and plates warm. Use of these should be carefully planned as keeping a cooked dish on hot refer to cooking after some time.

Using a Salver

1. Salver may be used to serve vegetables to clear glasses and to serve drinks and tea or the coffee.
2. Put a cloth on the salver to stop slipping and prevent conduction of heat, thus avoid burning of the palm of the hand.

Using a Tray

Restaurant do not carry bottles of wine or mineral waters on tray except in the case of wine bottles in cradles or wine coolers.

Trays are also used for cleaning dirties from the side stations and should always be loaded safely. Place the heaviest article in the middle to aid balance. Glasses should always be carried separately and not in the same tray as dishes. Apart from the risk of them being smeared with grease or sauce.

Carrying Trays

Continental trays are largely used for smaller orders in room service. Balance of these trays is very important and each how steward should always be trained perfectly by made to carry trays loaded with bricks, empty bottles, etc. Trays should always be carried by the centre and preferably with one had. A lighter tray is easily carried with the tips of the fingers and heavier with the flat open palm.

Clearing dishes and plates

Collect plates only when the guest indicate by putting fork and knife together across the plate. Remove all the plates after all the guests have finished, the host being in the end. Priority should be given to removing plates quickly to keep the table tidy and to also give an impression to the guest that he is not forgotten.

Clearing vacated tables, changing table cloth

A vacated table should be immediately cleared off the items. Replace crute sets, flowers and clean ashtrays immediately. Always change the table cloth after each service. Check that chairs are free from crumbs when a table is being cleaned.

Spills on the table

Water, wine, sauce or similar moist matter can be spilled on the cloth. For such spills, such as sauce, take a small plate with a knife or spoon to the table and remove the sauce by scraping. When the sauce is removed, cover the stain with a napkin or white paper.

Relating over during service

During service the cutlery should never be carried with the hand, even in a cloth, but always on a salver or large plate which has been covered with a napkin to deaden and noise. The fork and knives should be touched underneath at right angles on either side. This facilitates handling, alloys noise, avoids possible accident and looks neater.

Moving a laid table

When moving a laid table, fold back the hanging drop of the table cloth so you grasp the table itself thus avoiding creasing the cloth.

Progressing the meal

Always be completely aware of the table and keep an eye on the guest (without staring) so that the guest can attract attention readily. Seek a customer permission before removing crute sets or any other item. Always remember to :

1. Customer can reach required items like salt and pepper.
2. If butter dish on table should have contains butter.
3. Give the Breadrolls regularly.
4. Always keep an eye on guests who smoke as the faster a cigarette is let the better impression of quick service is created.

Accompaniments

Before the main dish make sure all the sauces and condiments are passed.

Finger Bowels

A finger bowel is served with certain dishes particularly food which a guest may handle such as asparagas or fresh fruit or curried dishes eaten with Indian bread. The finger bowl should be served after the meal, when the main dish is removed. After the dessert serve cold finger bowl.

Crumbling

Just after the meal and before the sweet remove the cruet set and also remove other remaining items, such as bottles. Always remove crumbs and bread from table before serving the sweet. For crumbling, use the neatly folded napkin or brush to bring crumbs to the table edge. From the edge they can be brought gently on to B&B (Bread and Butter) plate.

Cover for sweet service

After crumbling, adjust the dessert fork and spoon in cover but do no bring your arm in front of the guest.

Coffee or Tea

Coffee or tea should be served in coffee or tea pot, milk jug, and sugar on a salver. The cup is placed on the right of the guest in advance asking the guest.

1. Would you like your coffee with milk or cream?
2. How much sugar?

Bill

Bill should always be presented after a guest asks for it.

1. Check the bill completely before presentation.
2. The bill is presented in folder.
3. Return the bill and the change to the guest.
4. If the guest insists on having the bill removed do so graciously by saying "Thank you very much sir."

Accompaniments of Dishes

Dish	*Cover*	*Accompaniments*	*Remarks*
1	2	3	4
Poached Salmon (Saumon poche) (Served hot)	Fish knife and fork Hot fish plate	Hollandaise sauce Mousseline sauce	
(Served cold)	Fish knife and fork Cold fish plate	Sauce mayonnaise	
Mussels (Moules mariniere)	Soup plate Underplate Fish knife and fork Sweet spoon	Brown bread and butter Cayenne pepper	A soup plate is used here owing to the large amount of sauce offered
Cold Lobster (Hemard)	Fish knife and fork Lobster pick Spare plate for the shell Finger bowl containing luke warm water and a slice of lemon Spare serviette	Sauce mayonnaise	The lobster pick would be used to help extract the flesh from the claw of the lobster.
Fish (Fried)	Fish knife and fork Hot fish plate	Segments of lemon or slices of lemon with skin removed	The three sauce mentioned are all cold sauces with a mayonnaise base
Fish (Grilled)	As above	Lemon as above Cold sauces as above Hot sauces : béarnaise + yrolienne	

1	2	3	4
Fish (Poached)	As above	Lemon as above Cold sauces as above Hollandiase sauce Mousseline sauce Beurre fondu	
Shell fish Cocktail (Cocktail ds crevettes)	Teaspoon Oyster fork Shelfish cocktail holder Doily on a sideplate	Brown bread and butter	The + tableware required may be place either on the right and left of the cover or on the sideplate on either side of the shelfish cocktail holder.
Avocado (Avocado vinaigrette/aux crevettes)	Cold fish plate Teaspoon	Brown bread and butter	On occasions a special avocado dish is used, replacing the cold fish place.
Soup			
Minestornie	Soup spoon Soup plate Underplate	Grated Parmesan cheese Grilled flutes	Clear soup heavily garnished with assorted vegetables and spaghetti
Onion Soup (Soup á l' oignon)	Soup spoon Soup plate Underplate	Grated Parmesan cheese Grilled flutes	French onion soup
Consommé	Sweet spoon Consomme cup and saucer Fish plate	Depending on garnish	
Consommé En Geleé	As for consommé		Jellied clear soup

1	2	3	4
Bouillabaaisse	Soup plate Underplate Fish knife/fork Sweet spoon	Thin slices of French bread dipped in oil and grilled	'Bouillabaisse' a Meditteranean fish stew
Meat			
Raost Beef (Boeuf Roti)	Joint knife and fork Hot joint plate	French and English mustard Horseradish sauce Yorkshire pudding Roast gravy	
Roas Lamb (Agneau roti)	As above	Mint sauce Roast gravy	It is traditional in certain areas of the country to offer either mint sauce or red current jelly with roast lamb. On occasions both may be offered.
Roast Mutton	As above	Red current jelly (saddle or leg) Onion sauce (shoulder) Roast gravy	As with roas lamb
Roast (Porc roti)	As above	Sage and onion stuffing Apple sauce Roast gravy	
Paultry			
Chicken (Poulet roti)	Joint knife and fork Hot joint plate	Bread sauce, Roast gravy Prsley and +hyme stuffing Bacon rolls, Game chips, Watercress	

1	2	3	4
Duck (Caneton roti)	As above	Sage and onion stuffing Apple sauce, Water cress, Roast gravy	
Goose (Oie roti)	As above	Sage and onion stuffing Apple sauce Roast gravy	
Turkey (Dindonneau)	As above	Cranberry sauce, Bread sauce Chestnut stuffing, Chipolatas Game chips, Watercress, Roast gravy	
Bortsch	Soup plate Under plate Sweet spoon	Sour Cream Beetroot juice Bouchees filled with a duck paste served in the order listed	Bortsh is a dutch flavoured consomme. Garnish with duck, diced beef and turned vegetables. Although a consomme type soup, a soup plate is used as part of the large amount of garnish and accompaniments offered
Turtle Soup (Tortue vraie aux Xeres)	Consomme cup Consomme saucer Sweet spoon	Brown bread and butter segments of lemon Cheese straws Measure of sherry	The segment of lemon may be offered in a lemon press. The sherry would either be added in the kitchen immediately before serving or from the gueriodon at the table by heating the sherry in a ladle over a lamp

1	2	3	4
Farinaceous Dishes			
Spaghetti	Joint fork Sweet spoon Hot soup plate on an Underplate	Grated Parmesan	The joint fork is placed on the right hand side of the cover and the desert spoon on the left.
Fish Dishes			
Grilled Herring (Hareng grille')	Fish knife and fork Hot fish plate	Mustard sauce	
Asparagus (Asperges au beurre fondue)	asparagus rack grid-serviette Asparagus tongs: right of cover Hot/Cold joint plate Joint fork eggs and a spare serviette	If served hot hollandaise sauce or beurre fondue If served cold: sauce vinaigrette	The joint fork is placed upside down on the right of the cover so that the joint plate when placed on the fork is tilted from right to left. The asparagus when being served should have the tips on the left hand side of the plate, the lower part, so that when the accompanying Sauce is served it settles over the tips of asparagus. Asparagus may be served either as an hors-d' oeuvre substitute or as a separate vegetable course.
Globe Artichoke (Articha)	Joint fork on the right of cover Finger bowl Doily on an underplate Spare napkin Hot/ Cold fish plate as appropriate	If served hot : hollandaise sauce or beurre fondue If served cold: sauce vinaigrette	Globe artichoke may be served either hot or cold as an hors-d' oeuvre substitute or as a separate vegetable course.

1	2	3	4
Paté de Foei Gras	Sideknife Sweet fork Cold fish plate	Hot breakfast toast with crusts removed, cut into grainless and served in am napkin on a side plate	True plate is made from goose liver. Often seen on the menu as Pate' maison made according to the recipe of the 'house'
Snails (Les escargots)	Snail tongs : left of cover Spail fork : right of cover Snail dish : which is round and has 6-12 identations to hold the snail shells Doily on underplate	Brown bread and butter	Snails shells are placed in the snail dish with their openings suppermost to stop garlic butter running out.
Smoked Salmon (Saumon fume)	Fish knife and form Cold fish plate	Cayenne pepper Peppermill Segment of lemon brown bread and butter	If not pre-portioned in larder then the side of smoked salmon would be carved on the buffet
Caviar	Caviare or fish knife on right hand side of the cover Cold fish plate	Hot breakfast toast Butter Segments of lemon Finely chopped shallots Sieved had boiled yolk and white of egg	Size of portion here will be approximately 30 gms or 1 oz
Chilled Melon (Melon frappe)	Sweet spoon and fork Cold fish plate Teaspoon if Charentaise melon is offered	Ground ginger Castor sugar Ogen, Charentaise	Variety of melon normally offered would be Honeydew, Cantaloupe,

1	2	3	4
Fresh Frawns (Les creveites roses)	Wine goblet on doily on Underplate Fish knife and form Cold Fish plate Finger bowl as for gulls' eggs Spare serviette	Brown bread and butter Sauce mayonnaise	The portion of prawns is placed over the rim of the wine goblet which would be filled with crushed ice. For special presentations of this dish silver dishes of diminishing sizes are placed one in another to form a cascade. This is called les crevettes roses on cascade.
Grape Fruit Cocktail (Cocktail de pamplemousse)	Coupe Doily on sideplate Grapefruit spoon or	Castor sugar	Other fruit cocktails served as a grape cocktail, e.g. Florida cocktail
Tomato Juice (jus de tomate)	5 oz goolet/small tumbler or clut shaped glass. Doily on sideplate Teaspoon	Worcester sauce	Worcester sauce : bottle shaken, lid removed. Place on doily on sideplate at head of cover. Serve tomato juice chilled and decorate edge of glass with twist of lemon
Oysters (Les huitres)	Soup plate or welled silver dish filled with crushed ice and placed on underplate Oyster fork Finger bowl filled with lukeward water and slice of lemon Doily on Sideplate at the top left hand corner of the cover Spare serviette.	Cayenne pepper Peppermil Chilli Vinegar Oyster Tobasco sauce cruet Half a lemon Brown bread and butter	Oysters are normally offered 6 per portion. If another 6 are required then a fresh service would be laid, Oysters are served in the deep half of the shelf on the crushed ice surrounding the half a lemon.

Chapter 22

Laying the Table

1. Have the tables and chairs in position and see that the tables do not cobble.

2. See that the tables and chairs are cleaned thoroughly.

3. The table on which a table cloth is to be used should be covered with a baize cloth.

   Use of baize cloth :

   (a) It protects the customers wrist from the edge of the table.

   (b) If deadens the noise of cutlery crockery and glasses while placing them on the table.

   (c) It helps the table cloth to hang freely and firmly.

   (d) It protects the surface of the tables from spills and scratches.

4. Crumpled or soiled table clothes should never be used. Always see that the right side is up.

5. Crockery to be used must be spotless. Clean and the cutlery must be well polished. Chipped crockery should never be used.

6. Each cover should be well balanced and should never be over crowded wet cutlery and glasses should not to be used.

7. Place only required cutlery needed for meal. The sequence of spoon forks and knives should be from outside to inside in the order in which they are to be used.

8. Knives and spoons should be placed towards the right of the plate while forks are placed towards the left. If no knife is required then the fork is placed on the right of the plate for, e.g., serving oysters.

9. The silver for appetizers may be placed in the usual position on the table alongwith other cutlery or in certain cases it can be placed on the plate itself for, e.g., fruit cocktail and various juices.

10. Cutlery should be brought from the side board or side table or dummy waiter on a salver or an a large plate, but never in bare hands. It should be handled with a napkin or waiter cloth. While picking up the cutlery always pick it up from the handle and not from the eating part. Glasses should always be held from their bases.

11. The water tumble must be placed on tip of the large knife.

12. The butter dish should be placed on the tips of the fork alongwith a butter knife and under plate.

13. The napkin or servieter is placed in the centre of the cover or on the side plate.

14. All the cutlery and plates should be placed half an inch away from the edge of the Table.

15. The cover should be directly opposite to each other.

16. The self of pepper pots are placed above each cover or between two or four covers.

17. While placing the plates, glass, which bear monograms place them in such a position that the monogram faces the customer.

18. The cutlery should be on the table or on the side board but never in side the pocket.

19. See that the cutting edge of the knife is always toward the centre plate, and the edge if the side knife is away from the centre plate.

Note : *When a table is to be relaid, after the guest leaves, make a movement flow in which the steward should make the movement from table to side station only three times*

1. *Cleaning the table.*
2. *Changing the linen.*
3. *Relaying of cover.*

MIS-EN-PLACE OF THE RESTAURANT

Some people think that the only duty of the waiter is to serve food and beverages to the customer when he sits down on his station. It is quite true that his main job is to serve food but this service of food will not be done in an efficient manner unless and until the waiter has done some preparation before the guests arrive and the actual starts. This part of the waiters duty is known Mise-en-place. This part of waiters work entails :

1. Cleaning of furniture and fixtures.
2. Placing tables and chairs in position.
3. Cleaning and polishing silver.
4. Laying of table cloths and sideboard covers.
5. Cover of place setting on the tables.

Organisation of Mise-en-place

Maiter d' Hotel is responsible for the proper organisation mise-en-place.

1. He directs waiters on a duty roster system for getting the various jobs done e.g. one may be detailed to look after cruets another the glass and so on.
2. It is the head waiters responsibility to ensure that all equipment used is properly placed and everything is spotlessly clean.

Evening Mise-en-place

1. In restaurant where breakfast is not served or where a special breakfast room separate from the restaurant is available.

   (a) Restaurant tables are completely cleared.

   (b) Salt cellars and cruets set are emtied and clean and placed near the service door.

   (c) Mustard sauces pots to be emptied and washed.

   (d) Flowers are removed to the flower room, usually in the house-keeping department. Alternatively they may be kept in a cool place.

   (e) Soiled table cloths and linen are removed.

   (f) Carpet is vacuum cleaned or if the restaurant is not carpeted it must be thoroughly cleaned by washing.

   (g) Fresh linen is brought from the part stock maintained and kept in readiness at the restaurant for use in the morning.

2. In restaurants where breakfast is served in the same dining room which is used for the service of lunch and dinner.

   (a) Restaurant tables are completely cleared.

   (b) Cruets sets are cleaned.

   (c) Soiled table cloths and other soiled linen removed and stacked.

   (d) Carpet is vacuum cleaned and general dusting done.

   (e) Clean table cloths are laid and a breakfast cover is laid.

   For continental breakfast

   (a) A side plate and a side knife.

   (b) A butter dish and a butter knife on a B&B plate.

   (c) A tea saucer and a tea spoon.

(d) A sugar pot with sugar tongs.

(e) A cruet.

For English or full breakfast :

(a) A side plate with a side knife.

(b) A butter dish on B&B plate with butter knife.

(c) A tea saucer and a tea spoon.

(d) A sugar pot with tongs (for cube sugar) and a sugar pot with tea spoon (for grain sugar).

3. **Side board must be come with respect to :**

(a) Supply of all cutlery used on the table.

(b) Supply of all crockery used as under plates and for the service of cold preparations.

(c) Ample supply of service spoons and service forks,

(d) Napkins and a supply of all table linen.

(e) Salvers and trays.

(f) Ashtrays.

(g) Water tumblers.

(h) An empty water jug covered with a folded napkin.

(i) Fruit baskets completed with a decaratively done up napkin.

MORNING MISE-EN-PLACE

I. In restaurant where breakfast is served the following mise-en-place is done in extension to the work started the previous night :

(a) Doors and windows are thrown open for some time, curtains drawn out and the room allowed to breathe.

(b) Pots for preserves are filled and laid on the table with individual jam spoons.

(c) Napkins are laid in their position.

(d) Water jugs are filled with cold water.

(e) Fruit baskets are decaratively done up with fresh seasonal fruit.

II. When the restaurant serves no breakfast but serves lunch and dinner the mise-en-place starts at about 10.00 AM the mise-en-place consists of the following in extensions of the mise-en-place of the previous evening

(a) Put tables and chairs in order.

(b) All silverware, flat was used during service must be polished

(c) The cruet sets must be filled with salt and pepper, fresh mustard prepared and oil and vinegar must be checked in the cruet set.

(d) Necks and lips of the sauce bottles must be wiped clean.

(e) Linen which has been spoiled must be exchanged and fresh linen obtained from the linen room.

(f) Lay the table cloths and side board covers.

(g) In some establishments flower arrangements for use in dining rooms have to be prepared by the waiting staff. In any case, flower arrangements have to be produced from the house keeping department.

(h) After all preliminary work of menial natural has been completely done, it is time to lay covers on the table.

Chapter 23

Cooking

What is Cooking?

"Application and removal of heat from food to get the desired result is known as cooking."

Methods of cooking

1.	Steaming	water media	
2.	Boiling	water media	
3.	Roasting	fat	
4.	Poaching	water	
5.	Grilling	fat	Rare, medium, well done
6.	Frying	fat	a. Saute, b. Shallow, c. Deep
7.	Baking 250°F 502°F	Radiant heat	
8.	Broiling	Dry heat	
9.	Smoking	Smoke	
10.	Stewing	Water	
11.	Braising	Liquid	
12.	Poeling	Fat put in even in a closed vessel on chopped veg.	
13.	Microwave	Cooking food on a bed of veg & potatoes (Roasting + Stewing)	

The traditional and classical methods of cooking which are employed are as follows :

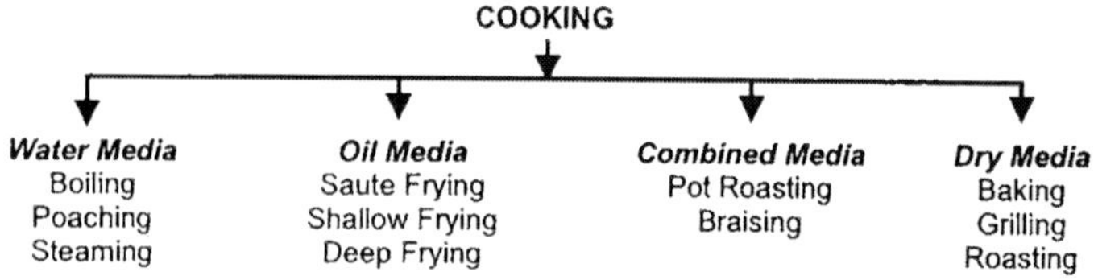

1. Water media

Boiling/Simmering

To cook in boiling water, stock or other liquid usually a lengthy processed used for cooking items, e.g. potatoes rice.

Poaching

To cook in water, maintained at temperature just below boiling point. A little acid in the form of lemon juice or vinegoris often added the liquid to help preserve the shape of the food used on delicate items of food, e.g., eggs, fish

Steaming

To cook in steam. The moist heat of food in its own juice with the addition of little liquid. After a lengthy process used to food cheaper, tougher items of meat, e.g., Irish stews, Hard fruit can also be stewed e.g. Stewed peas, apple or pineapple.

2. Oil Media

Deep frying

To cook by immersion in hot fat/oil. A quick method of cooking or reheating can be used for larger times of food which are often given a coating of butter to seal the food and enable it to cook in its own Juice e.g., fried fish, french fries.

Shallow frying

To cook by placing in a shallow layer of hot fat or oil. Usually a fairly quick method of cooking used for quite tender small or thin items of food. The fat must be at the correct temperature. Fat which is too hot will burn the food and cold fat will turn the food 'soggy' as the oil will soak into the food, e.g., fried eggs.

Saute frying

To 'jump or toss' small items of food in a pan containing a very little hot fat or oil thus preventing it from sticking to the bottom. Similar to shallow frying, often a sauce is made from the resulting liquid in the pan, e.g., Sauted mushrooms, Sauted Potato.

3. Dry heat

Baking

Cooking in the dry heat of an oven with heat applied from all sides, e.g. Bread.

Grilling

The item of food is cooked firstly on one side then turned over to allow the heat to penetrate quickly to the centre of the item. The heat is applied from one side, e.g. grilled lamb chops.

Roasting

Similar to grilling but reserved for larger items of high quality food which are allowed to cook more slowly than in grilling.

4. Combined Media

Pot Roasting

To cook by browning the item of food then placing in an enclosed pot with liquid and vegetables. This is done for fairly good quality food items

Braising

The item of food is placed on a baked of roughly chopped vegetables and half covered in a flavoured cooking liquor, the vapour from which helps to cook and flavour the food. This is later finished and turned into a sauce to a company the food.

Other common cooking terms

These terms apply especially to grilling

Rate	-	Very undone
Medium	-	Just done
Well-done	-	Well cooked

Fume	-	Smoked
al' anglaise	-	in the English style
en Aspic	-	in aspic jelly (Cold)
		a fireproof dish and served in the same.
farei	-	stuffed
garni	-	garnished
froid	-	cold
au gratin	-	topped with grated cheese and crumbs and grilled
glace	-	iced

Chapter 24

Stocks

Definition

It is the solution extract from marrow of mutton, fish, poultry, veal, beef game or vegetables in water and mirepoix terms.

Mirepoix

Root vegetables cooked together.

Roux

Butter oil and flour in half and half quantity and cooked together.

Types of Roux

White, Golden, Brown.

Beurre Manie

Butter and flour in half and half quantity it should be uncooked.

Bouquet Garni

Bouquet Garni is the mixture of herbs (Bayleaf, Thyme, Peppercorns and Parsley) rap in mouslin cloth and put into, during cooking, to get the good flavour.

Chapter 25

Soup

CLASSIFICATION OF SOUPS

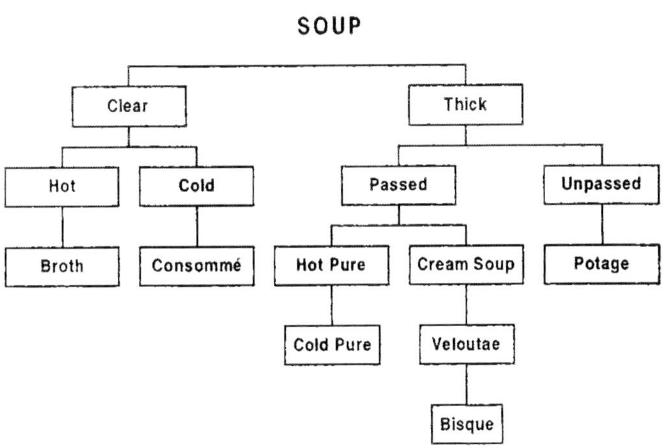

DEFINITION

A liquid food consisting of meat, seafood, poultry, cereals or vegetable served clear, thick, hot or cold mainly at the beginning of the meal of enhance appetite.

1. Consommé

Clarified by adding meat and egg white to the cold stock then simmer for 45 minutes then strained.

2. Cold consommé

Cook-Flavour it-garnish it chilled it-becomes jelly if less add gelatin.

3. Bouillon

Rich consommé

4. Puree

Vegetable based - Veg is sauted-stock added-thickened (with rice/cereals)-strained, e.g. puree perineometer.

5. Potage

Veg based, e.g., potage st. Gemain (Green Peal)

6. Cream Soup

It is made by combination of stock + Roux + Seasonings, e.g. cream of chicken.

7. Bisque

A thick shellfish soup, e.g., Bisquede Homaret.

8. Cold Soup

These are served chilled, e.g., vichyssoise-chilled potato and leek soup. First cooked together and chilled; Gazpacho-Fresh tomato and capsicum blended with paprika powder garnished with finally. Chopped capsicum, carrots and sour cream.

Creme-Doria-Cucumber blended with cream.

Chapter 26

Sauces

A sauce is normally a rich flavoured, thickened stock or liquid used to give moistness or flavour or otherwise to accompaniment food.

Quality points to note in sauces are :

1. There should be no taste of uncooked starch.
2. The texture must be smooth and velvety.
3. Sauces should never be paste-like to excessively thick though the consistence varies from sauce to sauce.
4. Sauces must be clear and have a nice seen.
5. There should not be any prominent flavour. It must be a blend of several suitable flavours.

Importance of sauces

Sauces are fundamental to continental cooking and play a big role in the final dish presentation to guest. Some important points to be known about a sauce are.

1. They are very nutrition, being rich in fats, starch and proteins.
2. Some sauces help in digestion, e.g., mint sauce.
3. They adds flavour to the food.
4. They moistness the food, e.g., white sauce added to vegetable.

5. They gives colour to food.

6. They can be the foundation of a dish, e.g. white sauce of cheese souffle.

There are six principal sauces from which other sauces are divided. These sauces are known as Basic or Mother sauces. They are :

Types of Sauces

1. Mayonnaise
2. Hollandaese
3. Bechamel (White Sauce)
4. Espaginol (Brown Sauce)
5. Veloute
6. Tomato Sauce

Sauces are made from

1.	Made from roux	- Bechamel, Espaginol, Veloute, Tomato
2.	Made from egg	- Mayonnaise, Hollandaese
3.	Made from custard powder	- Conflower, Sweet sauces, Custard sauce
4.	Uncooked sauce	- Mint sauce, Mayonnaise
5.	Pureesauces	- Tomato sauce, Apple sauce
6.	Fruit syrup & jam	- Strawberry jam sauces, camberry sauce.
7.	Miscellaneous sauces (Indian Sauces)	- Gravy, Curry.

Mother Sauces	Ingredient	Derivatives	Ingredient of Derivatives
1	2	3	4
Maoyonnaise	Salad Oil, Seasonings Emulation of egg yolk and Butter with and pepper touch of Lemon	1. Tartar	Shallots, may, chappers, gherkin, terragon leaves.
		2. Cocktail	May, tomato, ketchup, cream, brandy
		3. Thousand inland	Tartar, cocktail (1/2 half)
		4. Remoulade	May, Gherkins, Carpers, parsely
		5. Gribiche	Tarar, hand boiled egg, chopped, green herbs
		6. Cumberiand	May, shallaots, orange, lemon (rijon), mustard, red current jelly, lime juice,
		7. Maltaise	Orange juice, orange peel, May
Espaynginole (Brown Souce)	Brown stock Brown roux Mirepoix	1. Bordelaise	B/sauce, shallaots, beef marrow, red wine, pepper, thyme, bayleaf.
		2. Chasseur or munte sauce	B/sauce, shallaots, white wine, butter, mushroom, tomato, tarragoen.
		3. Pepper	Demiglace, shallaot, crushed pepper, brandy, seasining. (Demiglance is a reduced basic brown sauce).

1	2	3	4
		4. Diable	Tomato flavoured demiglace, pepper, white wine, vinegar, thyme, shallaots, bayleaf
		5. Robert	Demiglace, shallots, butter, white wine, mustard
		6. Mederia	Espagnole, Madiera wine.
		7. Zingara	Demiglace, Tomato Soup, mushrooms, julliene of ham, pickled tounge, truffle, sea soned with paprika.
Bahcamel (White sauce)	roux, milk bouquet garni mirepoix	1. Mornay	Bechamel, butter, grated cheese, egg yolks.
		2. Soubise	Bechamel, onion puree, nutmed
		3. Cardinal	Bechamel, cream, fish stock, lobster, butter cayenne pepper, truflles
		4. Egg sauce	Bechamel, hard boiled eggs, cream
		5. Curry	Bechamel, curry powder
		6. Nantu	Bechamel, cray fish, cream, butter, brandy, cayenne pepper
Tomato Sauce	Tomato puree, white stock, butter pepper, garlic	1. Brubcue	Tomato sauce, paprika powder
		2. Brettone	Tomato sauce, chopped fried onion, white wine, parsley

1	2	3	4
		3. Provencale	Tomato sauce, concasse, sugar, parsly, extra garlic, onions, white wine, veal stock
Velouté sauce	White stock golden roux	1. Berry	Veloute, shallots, butter, wine, fish, fumet, parsely, (fumet is essence of fish)
		2. Allemande	Veal Velute, egg yolk, butter
		3. Supreme	Chicken velouts, cream, butter
		4. Tamor	Fish veluté, oil, seg of lemon, paprika
		5. Nuberry	Fish velouté, egg yolk, sheery
		6. Amandine	Fish velouté, butter, almondflakes
		7. Ivory	Supreme, beam glaze
		8. Normande	Fish velouté, oyster, mushroom, egg york, cream and butter
		9. Sauce vin	White wine sauce-fish veloute, white wine, blane, egg yolks, butter, lemon
Hollandaese	Egg yolk, melted butter, seasonings, lemon, pepper, mixed on a double bowler	1. Bearnaise	Hollandaese, tarragon, vinegar
		2. Mausselline	Hollandaese, whipped cream
		3. Capre	Hollandaese, capers
		4. Maltalse	Hollandaese, orange juice, orange jelly
		5. Choron	Bearnaise, Tomato Puree.

Dressings

French	1:3 vinegar to oil, S&P and little french mustard
Vinaigrette	2:1 vinegar to oil
American Vinaigrette	1:1 vinegar to oil and seasoning
Italian	1:3 vinegar to oil, garlic (finely chopped, oregano, S&P)
Sherry wine	Sherry loine vienegar 1 cup, salad oil 2 cups, egg yolks, S&P
Russian	Maoyonnaise, chopped be etroot, touch of caviar.
Ravigeette	Vinegar, green herbs

Chapter 27

Meatcuts

Chicken cuts

English	-	*French*
Wing	-	Aile
Winglets	-	Ailerous
Drumstick	-	Pilonde cuisse
Thigh	-	Grasde causse
Breast	-	Poitrine
Carcess		

Fish cuts

Darne	-	Cut across the bone
Fillet	-	Large noise cut ebones deskinned
Goujon	-	Fish Finger
Paupietts	-	Roll of fillet

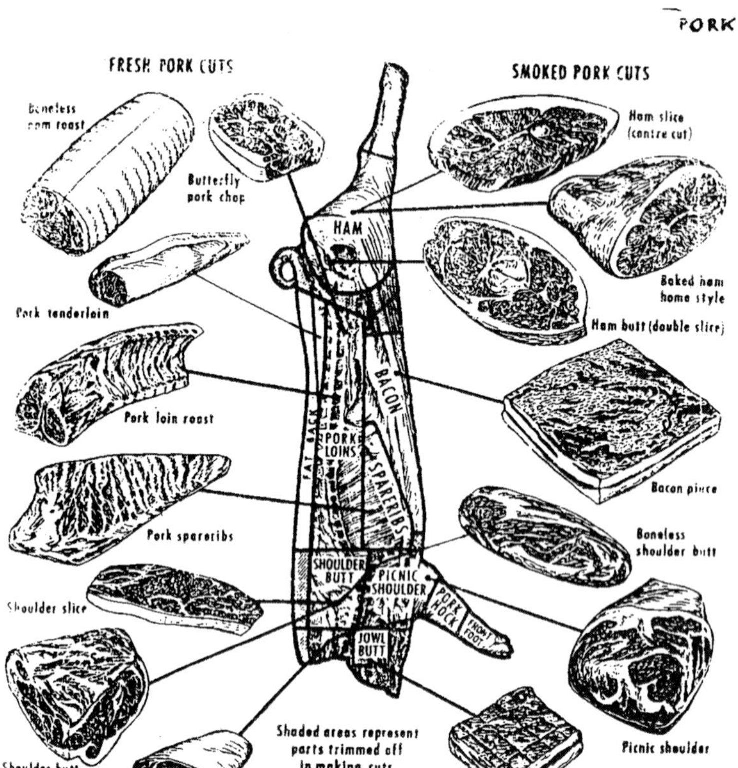

American cuts of Pork

Figure

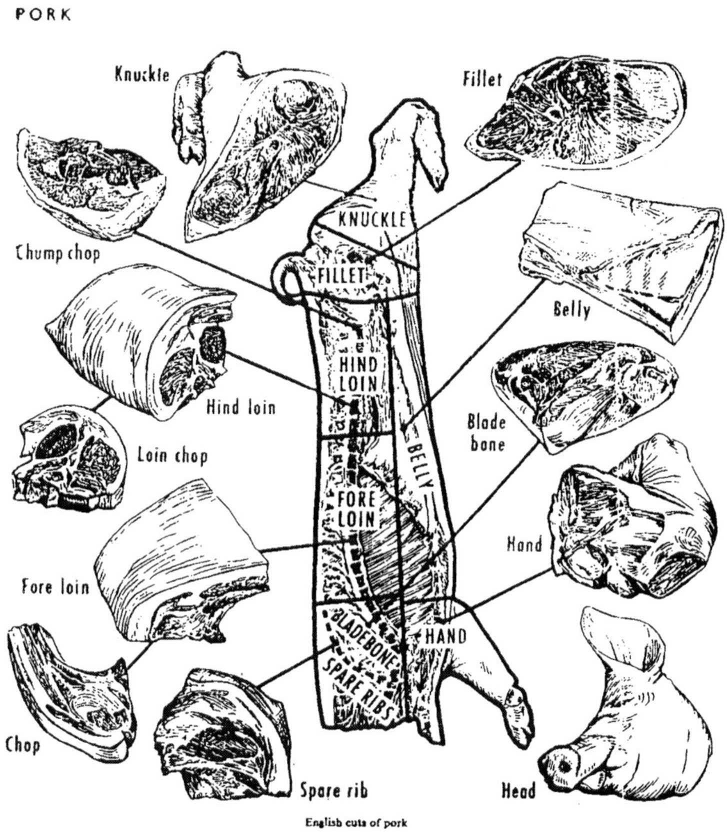

English cuts of pork

Figure

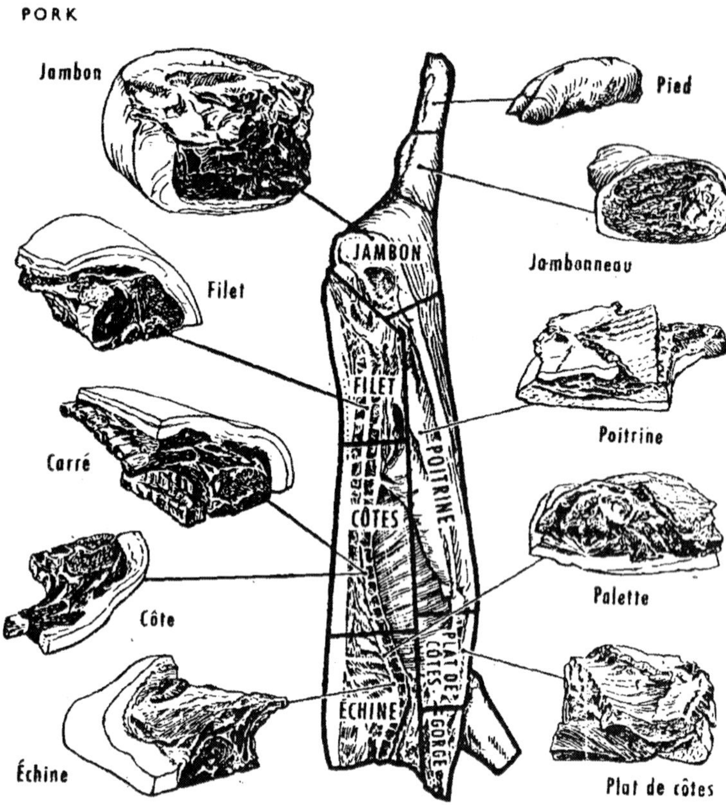

French cuts of pork

Figure

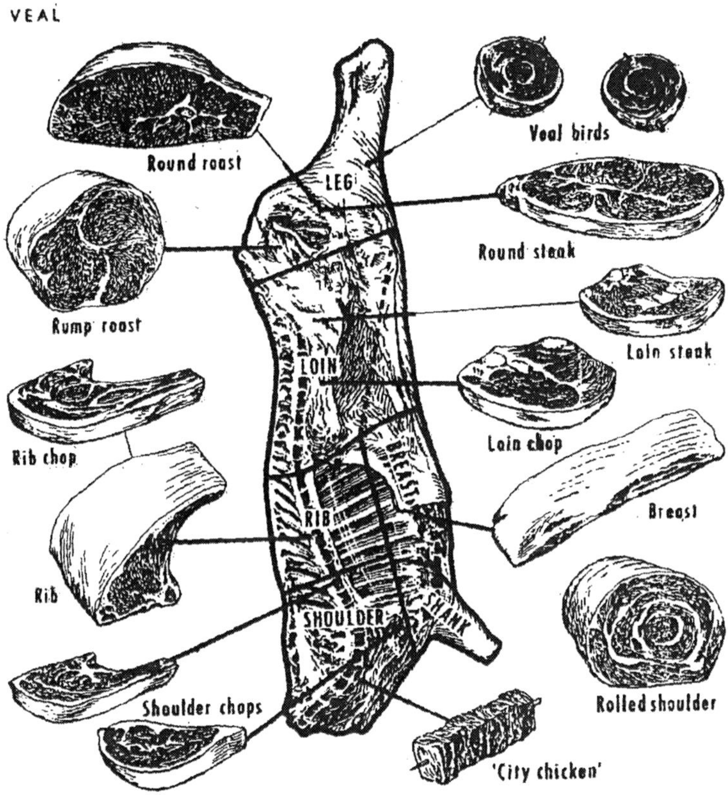

American cuts of veal

Figure

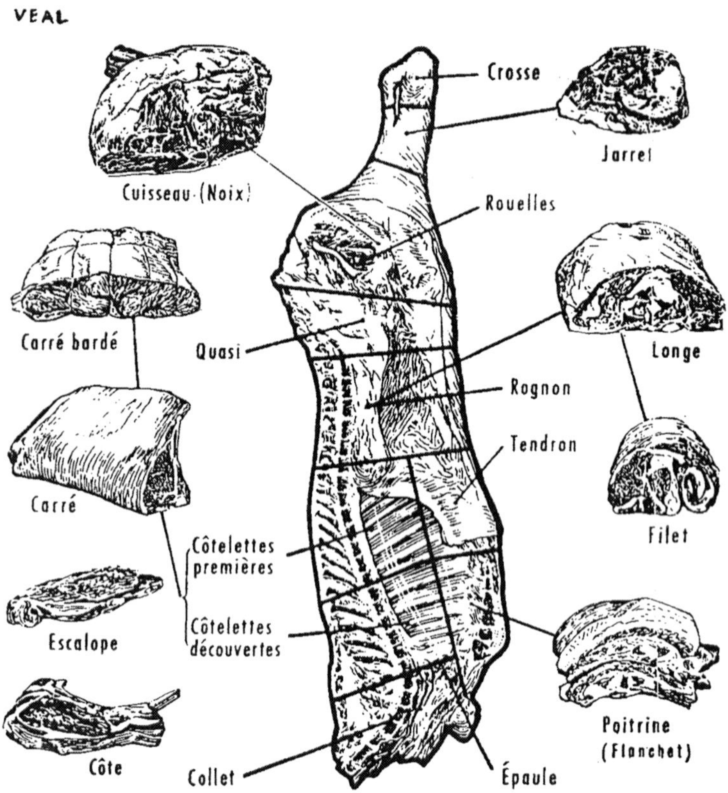

French cuts of veal

Figure

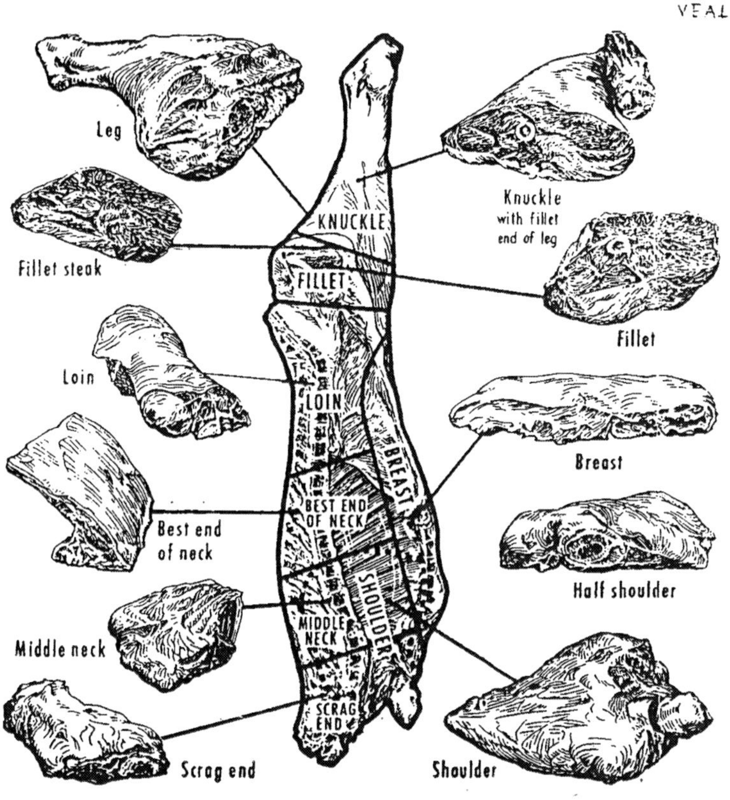

English cuts of veal

Figure

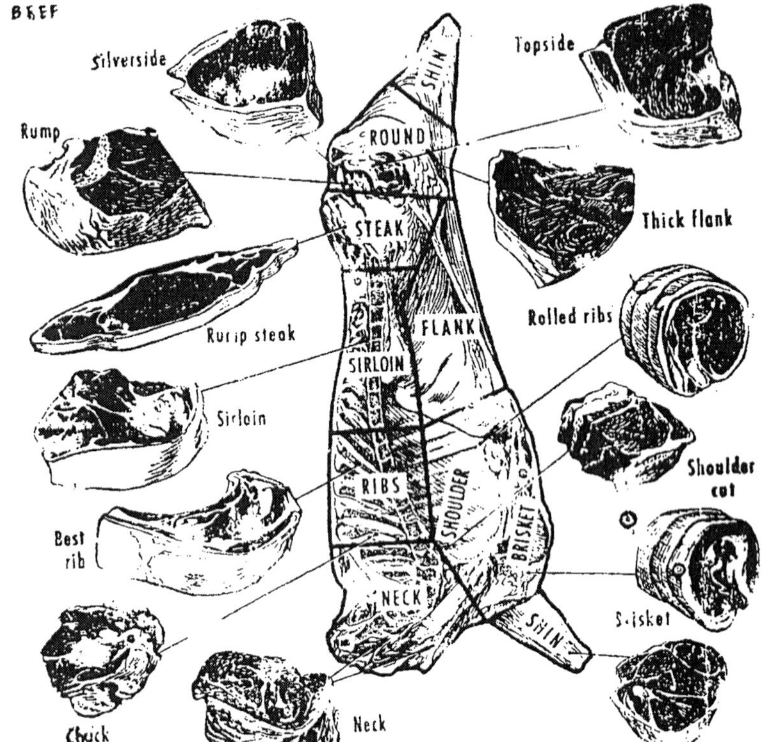

Chart of cuts of beef, English style

Figure

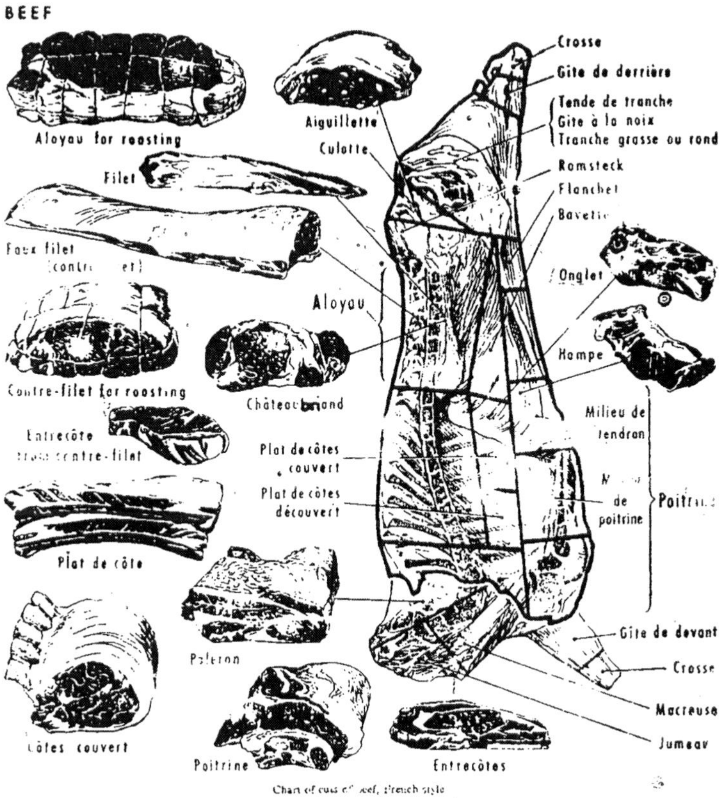
BEEF
Crosse
Gite de derrière
Tende de tranche
Gite à la noix
Tranche grasse ou rond
Romsteck
Aiguillette
Culotte
Aloyau for roasting
Filet
Faux filet
Aloyau
Onglet
Hampe
Contre-filet for roasting
Châteaubriand
Entrecôte
Milieu de tendron
Plat de côtes couvert
Plat de côtes découvert
de poitrine
Plat de côte
Paleron
Gite de devant
Crosse
Macreuse
Jumeau
Côtes couvert
Poitrine
Entrecôtes

Figure

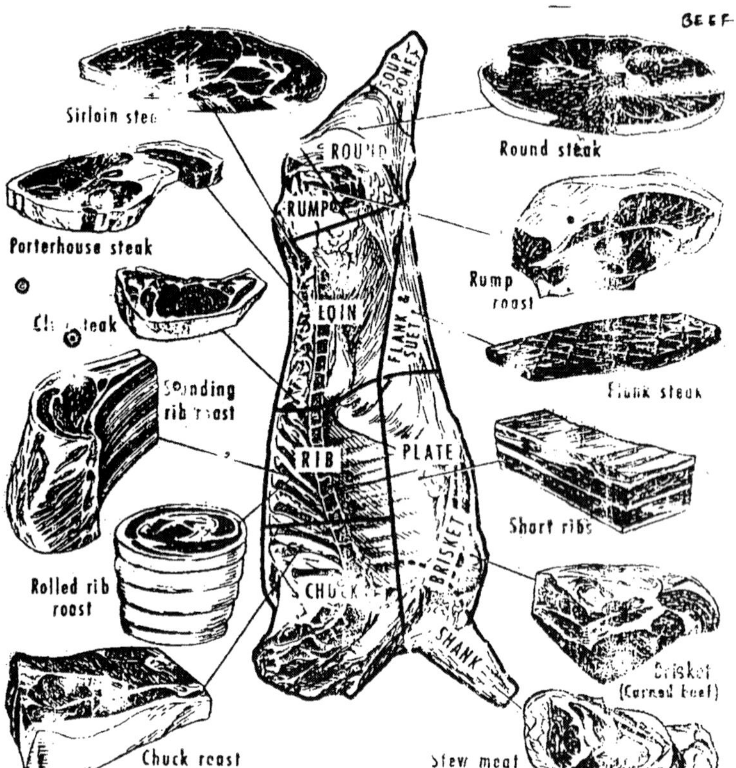

Chart of cuts of beef, Ameri…

Figure

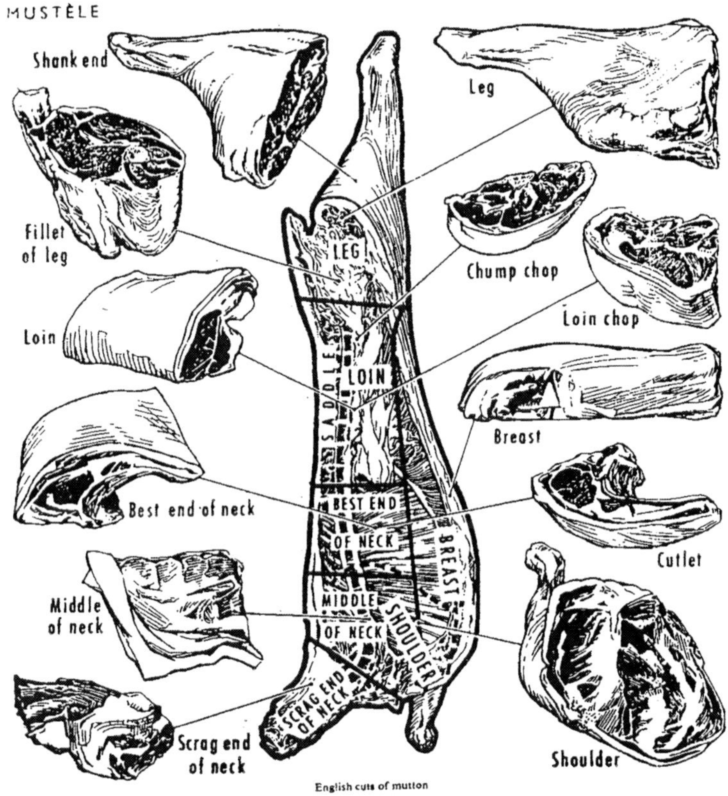

English cuts of mutton

Figure

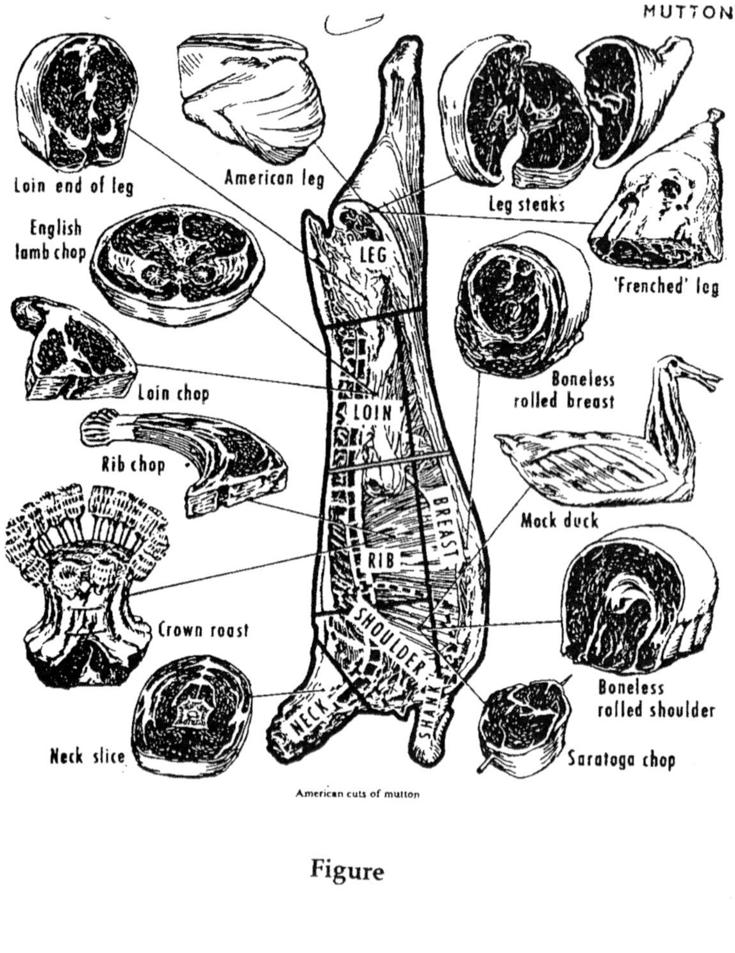

American cuts of mutton

Figure

MUTTON

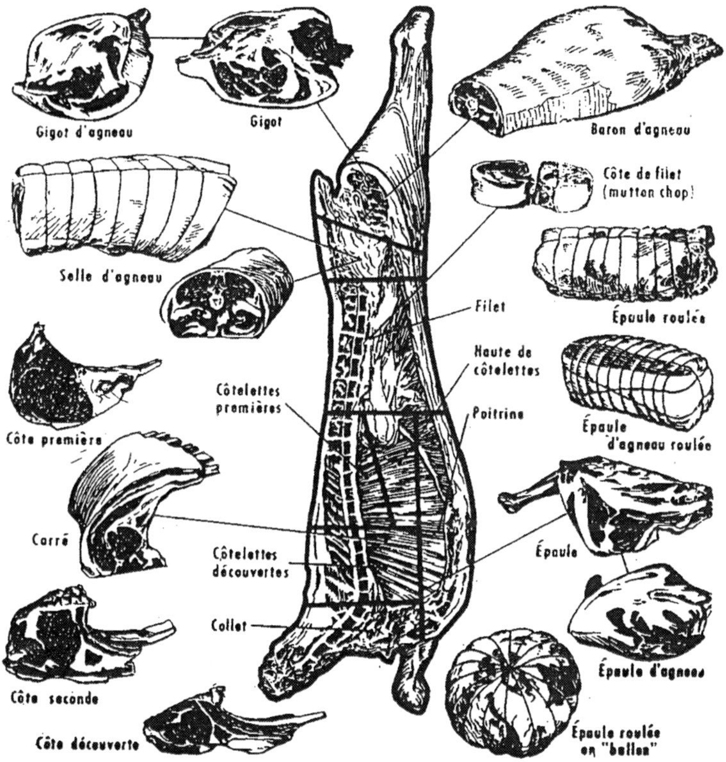

French cuts of mutton

Figure

Chapter 28

Cheese

It is the bio-product of milk, it can obtained from milk or cream or together it all depend on the kind of the cheese. It is coagulated with rennet (Rennet is obtained from the intestine of calf) or acetic acid. Cheese is made from cows, goats milk.

Different kind of heat and different degrees of pressure makes different kind of cheese. In some kind of cheese are with herbs or spices but it all depend on the kind of milk used. Curd id is fermented cooked.

The light pressure will give light cheese and the hard pressure will give hard cheese. Cheese can be softer then cream and cheese can harder like rock.

Soft or Cream cheese : Whole milk or enriched milk (cream added) is used.

Semi hard: Part whole milk and part skimmed (Fat removed).

Hard : Skimmed Milk.

Other kinds of cheese are :

Blue : Shewered with copper wires with a kind of penicillin (Penicillin glaucum introduced).

Processed : Made by grinding the cheese and then cooking in vacuum cookers, pouring into moulds and wrapping in tin foil.

The cheese is rich in proteins, fats, vitamins A, B^1, B^2, C and mineral salts.

Renet : Rennet is aptend from the intestine of calf.

Process of cheese making

1. Milk is pasteurised and curdled by the natural fermenters in milk or by adding 'starter' bacteria.
2. Coagulation is induced by adding 'renned' an extract from the abdomen of young minerals. Proteins in the curd solidify readily and leave behind the watery part called whey.
3. Drainage is the stage when all moisture is allowed to escape by pressing between boards, soft cheese is gently pressed whereas great pressure is applied in case of hard cheese.
4. Moulding is when the curds are put into moulds and allowed to mature; fermentation occurs and the rind flavour and its taste develop.

Types of cheese used in kitchen: Parmesan (Italy), Gruyere (Switzerland), Cheddar (England) and Amul Cheese (Indian) when cooking cheese it should not be over cooked as it is liable to become very indigestible and is liable to 'split' (the fat separating from each ingredients). It should always be combined with starchy food as it aids the digestion. Cheese also be comes rubbery after over cooking. When storing cheese. It should be kept in a cool dark place and should be wrapped in grease proof paper to prevent drying. If it is kept at a temperature which is too cold, it will sour and if put in a warm place it will sweet. It should he cold stored always away from food or it will absorb flavour.

International known Cheese

Cheese	Origin
Bel paese	Italy
Brie	France
Pelite suisse	Swiss
Camembert	France
Gorgonzola	Italy

Parmeson	Italy
Emmenthol	Swiss
Gruyere	Swiss

Description

1. Mild Italian Cheese : Large round and flat, creamy colour, texture soft, mild sweet flavour, ripens quickly.
2. Soft texture, colour is reddish brown crust, yellow paste.
3. White colour, creamy texture prepared from unsalted fresh cream.
4. Colour and texture soft and creamy.
5. Semi-hard blue mould cheese, prepared from cow's milk.
6. A hard white cheese. A kitchen cheese takes 2 to 3 years to mature.

In colour it is ivory to light yellow waxy texture. Has holes approximate to the size of cherry.

As above, but firmer in texture, holes are smaller takes longer to nature than emmentinol.

Cheese	*Origin*
Gouda	Dutch (Netherlands)
Edam	Dutch
Roquefort	France

English Cheese

Cheddar	England
Chesshire	England
Stillon	England
Lancashire	England

Description

Dutch cheese pale yellow in colour, waxy texture.

Dutch cheese red ringl, yellow in colour. Texture waxy round like.

Blue mould cheese similar to geogonzola. Has a semi-hard texture.

Colour yellow, texture close. Clean and mellow nutly flavour.

White or orange on colour, texture crumbly. A strong flavoured cheese.

Finest English table cheese. It is a hard blue mould texture flaky. It is usually spooned out for service.

It is of a mild flavour, crumbly texture and when fully nature can be spread like butter.

Cheese	*Origin*
Wensleydale	Yorkshire
Carerphilly	Wales
Deby	England
Double	England

Gloucester

Other brands available in our hotels are kraft, a processed cheese from Australia and Amul, an Indian processed cheese.

Description

It has a flaky texture and is pale in colour. It has a 'honeyed' after taste.

Creamy white of colour, semi smooth texture and delicate flavour.

Close texture, pale honey yellow colour mild when young but when fully maturated has much fuller flavour, some times flavoured soge.

Corn coloured; velvety textured similar to cheddar in taste.

Service of Cheese

The cheese course 'SAVOUREX' precedes a dessert course (fresh or dry fruits) but follows an Entrement or a sweet item.

- A B&B (Bread & Butter) knife and frehs fork, with a B/B plate are used as 'cover'. Accompaniments are placed on the table after laying the cover. They are :

- Salt and pepper
- Butter
- Celery, radish (if in season)
- Fennel bulbs (belongs to the onion family) are put on a bed of crushed ice in a Rolly Polly glass or a champagne saucer on an underliner.
- Assorted cheese; biscuits cream crackers, brown bread in a bread boat.
- If served in a buffet, pickled vegetables and olives are placed in a supreme bowl in a glass insert.

The cheese board of trolley is presented to the guest, the selected variety is cut with a cheese knife and cheese is picked up in the teeth of the knife. Never remove the rind of cheese while cutting it. An assorted plater of cheese is served on dessert plate with a brass plate as underliner. The rind is cut in this case.

The best wines with cheese are red such as Beaujolais and chateau Neuf dué Pape.

Chapter 29

Hygiene Rules

Hand must be washed

1. On arrival of work.
2. When returning to work area fallowing a meal, cigarette or coffee break and after going to the toilet.
   (a) When working with a food group and changing to another food group, e.g. changing from raw to cooked meat or changing from vegetables to bread/dairy etc.
   (b) Any time your hands feel unclean/dirty.
3. Always wear gloves when preparing food that has no/no further cooking, e.g. salads, rolls etc.
4. Change gloves as frequently as hands are washed.
5. Soak knives in clorex solution at night every day.
6. Chopping boards must go through the dishwasher when :
   (a) Changing food groups
   (b) Going on a break or off duty.
7. Wash down work surface with hot water, detergent and a clean cloth before going on a break or off duty.
8. Change J-cloths regularly.

Food Group

There are four main food groups.

1. Raw meat/fish
2. Cooked meat
3. Vegetables (including salads)
4. Bread and dairy.

Cross-contamination (mixing) of these food groups be prevented.

Reducing risk of Causing cross contamination

(a) All opened cons/packets etc. must be put into containers and covered (with e.g. tinfoil) and have the name of the item and data written on top.

(b) Food must be stored under designated area of fridge/freezer.

(c) Check expiry dates before use.

(d) Rotate old with new stock.

(e) Fruit and vegetables (including salads) must be washed and soaked in a dorox solution for 5 minutes and rinsed well before use.

(f) Only cook food 1/2 hour before service never save and reheat.

(g) Cook all food as close to service as possible and keep hot in a bain marie.

(h) Temperature reporting of equipment.

Fridges —	Fridges 18°C
	Freezer 5°C
Dishwasher	Wash 60°C
	Rinse 70°C
	Final rinse 82°C

HAND WASHING PROCEDURE

1. Use soap and warm running water.
2. Rub your hands vigorously.
3. Wash all surfaces, including: backs of hands, wrists, between fingers and under fingernails.
4. Wash and brush each finger 30 seconds, use a nail brush.
5. Rinse well in running water.
6. Dry hands with a paper towel.
7. Shut off the water using a paper towel instead of bare hands. Open restroom doors with paper towel, then discard paper towel.

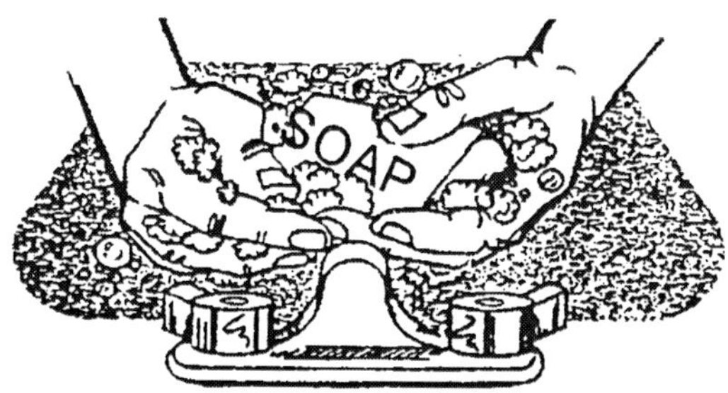

WANTED
• HAIR CLEAN AND COVERED
• GOOD HEALTH
• CLEAN UNIFORM
• NO JEWELLRY
• HANDS CLEAN AND SANITARY
• DOESN'T SMOKE ON THE JOB
SAFE FOOD
HANDLERS

Figure

CROSS CONTAMINATION
UTENSILS
Wash and sanitize after every use
CUTTING BOARDS
Wash and sanitize after every task
HANDS
Wash frequently and thoroughly
EQUIPMENT
Clean and sanitize after every use
PREVENT A DANGEROUS RECIPE

Figure

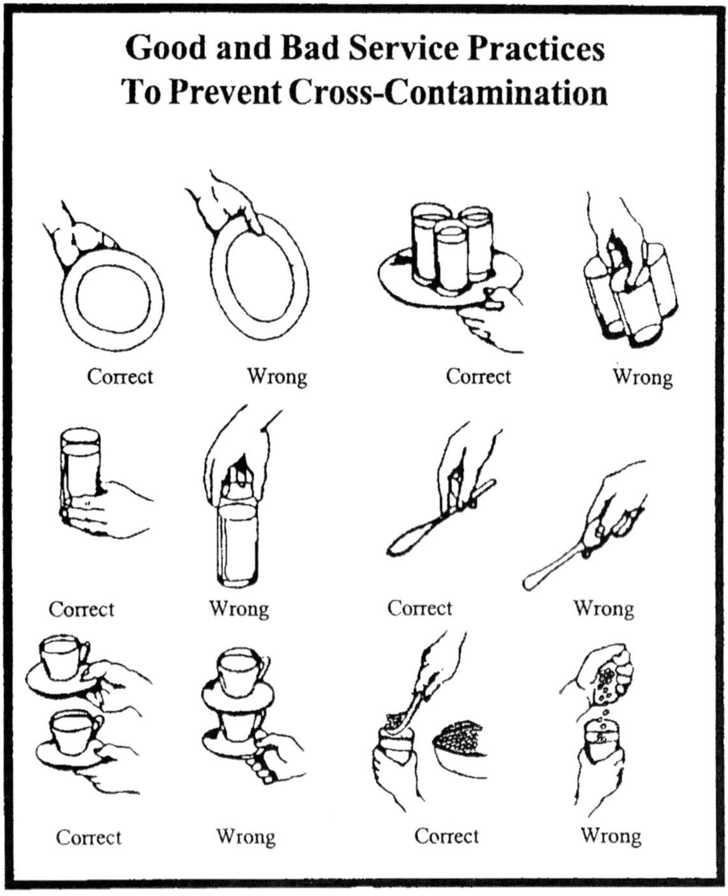
Good and Bad Service Practices
To Prevent Cross-Contamination
Correct
Wrong
Correct
Wrong
Correct
Wrong
Correct
Wrong
Correct
Wrong
Correct
Wrong

Figure

WASH
YOUR HANDS
AFTER
Touching any soiled object or surface, soiled clothing, etc.
Using a handkerchief or tissue.
TISSUE
Cleaning, taking out the garbage, or putting away supplies.
Flour
Sugar
Smoking, eating, or drinking.
Touching or scratching any area of the body (ears, mouth, nose, hair, etc.)
Handling raw food - particularly meat, poultry, and foods served uncooked.
Handling dirty dishes and before handling clean dishes.
Going to the restroom

Figure

CORRECT TABLE AND KITCHENWARE WASHING BY HAND

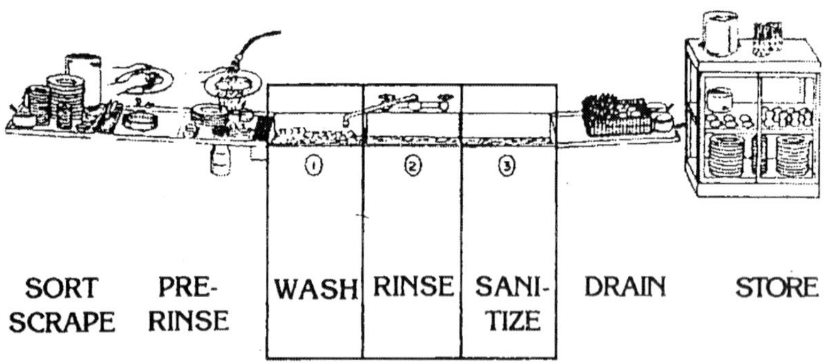

Figure

1

WASH
WITH A GOOD DETERGENT IN WATER AT NOT LESS THAN 110°F

2

RINSE
IN CLEAN

HOT WATER

3

SANITIZE
IN WATER AT NOT LESS THAN 170°F (OR) ADEQUATE CHEMICAL SOLUTION FOR 1 MINUTE

	(PPM) MIN-MAX
CHLORINE	50-200
Iodine	12.5 - 25
Quat. Amm.	100 - 200

Figure

Chapter 30

Bacteria

Bacteria are uni-cellular macro-organisms which can only be seen under a microscope. They are everywhere in our surroundings and as they cannot move by themselves, they are transferred to something by coming into direct contact with it.

The bacteria cell shape can be

1. Spherical known as COCCI.
2. Cylindrical rodlike termed as Bacillar.
3. Spoilage - which decompose food.
4. Pathogenic - which cause illness.

Some bacteria form spores which are bodies that can survive adverse conditions for long periods of time. During this period they do not multiply but on return to favourable conditions they become normal bacteria again and then multiply.

Some bacteria strains produce poisonous substances within themselves or in the food medium, which when consumed cause food poisoning. These toxins may be indotoxins, exotoxins or entero-toxin.

Others cause infection by virtue of large number of them in food entering the digestive system multiplying further Bacteria multiply further. Bacteria multiply by binary fission, i.e. division of one cell in to two, this takes place, every 20 minutes under suitable conditions.

Therefore, one bacteria could multiply in 10-20 hours to between 500 million and 1000 million bacteria.

Favourable, unfavourable conditions for their growth

1. *Food* : Bacteria require food for their growth. They by Antotrophic which can manufacture their own food or Heterotrophic which one on organic matter and hence be of concern to us. Out of the Heterotrophic groups few are sopuogur which feed on dead organic matter whereas others more harmful and disease causing ate pathogens which feed on the living organic matter, i.e. even within us. However, in general all strains of bacteria prefer food with higher pH* value, i.e. non-acidic food, e.g. milk, cream, cheese, meats, especially ham, pork, fish and other such foods that have ingredients more favourable for bacteria growth. They are referred to as potentially hazardous foods.

2. *Nutritients* : Food consumed gives nutritients to bacteria for their growth. They require mainly carbohydrates, proteins and few mineral and vitamins.

3. *Water and humidity* : Bacteria required water to carry on their activities, may it be their own growth and reproduction on food spoilage. Only foods that have some moisture can be acted upon by them. Dehydration kill them in most of the cases. Though a few spore forming strain survive that conditions and become active again the moment water is added in the medium. A food medium should have minimum 15 per cent moisture available to be liable to bacterial attach, otherwise its safe.

4. *Temperature* : This is the most important conditions essential for the growth of bacteria. All bacteria grow well at room temperature. However, a few strains may like cooler or higher temperatures. There are three types of bacteria classified according to their temperature requirements.

   (a) Phychrophilic - cold loving bacteria 45° - 55°F

   (b) Mesophillic - moderate heat loving 85° - 105°F

   (c) Thermophillic - heat loving bacteria 130° - 140°F

*pH : potentially hazardous

Since not even cold loving or heat loving bacteria grow below or above 45°F and 140°F respectively the some is regarded to as the danger zone for all foods.

Or if taken into account for all bacteria in general the following chart gives idea of their growth rate at different temperatures.

Chart of growth of Bacteria in different temperatures.

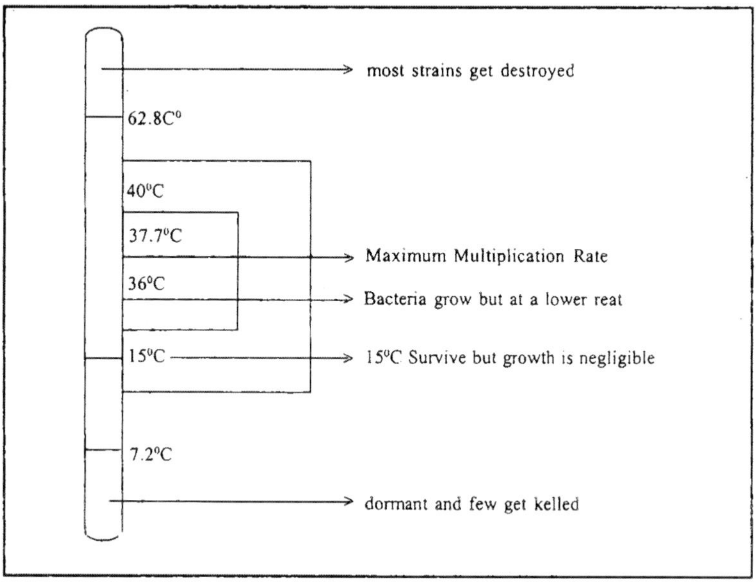

100°C

5. *Air* : Another requisite for Bacteria growth and multiplication is proper atmospheric conditions. Most bacteria require oxygen for their growth and are known as Aerobes. Whereas there are a few stains which can grow only in the absence of oxygen known as Anaerobes. In adverse conditions both aerobes and anaerobes get either killed or remain dormant till conditions reverse.

6. *Light* : Bacteria thrive well only in conditions or lack of absence of light. Dark, dingy, humid corners increase their rate of multiplication. Strong rays or sunlight, i.e. ultra-

violet rays, prevent their growth and even kill some. This fact is made use of in water tanks and swimming pool sanitation.

7. *Osmotic Pressure* : Too much variation in the densities of cell cytoplasm and that of the food medium, results in the ruptuing of the bacteria death, this is made use of in the preservation of Jams, marmalades, etc.

8. *Time* : Last of all, having enough time to grow and multiply is essential for bacteria to be able to reach the number which is dangerous to us. It depends on the presence of all the other optimum conditions and also on the initial number of bacteria introduced in the food. At an average it requires 5-6 hours for a food to be so contaminated that it is proves harmful to us.

Chapter 31

Tobacco

Tobacco is grown in tropical or semi-tropical climates in countries such as Cuba, Jamaica, Turkey, Rhodesia, Java, Sumatra, India, Syria and parts of the USA. The yellowish green leaves of the plants are harvested wilted, dried and cured after which they are ready to be used by the cigarette or cigar manufacturers.

Curing is of two types

Fermentative Method : Where the leaves are hung in a shed and heat treated at 170°F for 4-5 days.

Non Fermentative Method : Where they are sun dried in roofless chamber. Virginia latakia, Periquee, Turkish and Egyption are some of the kinds of Tobacco leaves.

Cigarettes

The dried tobacco leaves are shredded by machines, the moisture is removed and the shreds are cut and roll allowed to mature. Machine roll the paper, pack the tobacco, gum the paper ends and cut the rolls into required sizes. The cigarettes are dried for 24 hours and then mechanically packed.

Serving oc Cigarettes

When a guest has ordered a certain brand of cigarettes offer it on a B/B plate with a doily paper. The packet should be opened with a couple of cigarettes pulled partially out of the packet. A matchbox is placed beside the cigarette pocket. Be available to hold up the flame to the guest cigarette. Do not put out the flame by blowing it but extinguish it by as with buds downward swish to the hand. Ensure that too many bud stubs do not collect in the ashtray and frequently replace with fresh ashtray. This is done by covering a full ashtray with an upside-down empty one and removing from the table.

Cigars

The best cigars in the world come from Havāña in Cuba though they are also made in Jamaica, Mexico, Burma, India, Holland and Russia. Hand made cigars are preferred to machine ones. A cigar is composed of three parts.

1. *Filler* : inner core of the cigar.
2. *Binder :* which holds the 'filler' together. The filler and the binder are together the 'Punch'.
3. *Wrapper* : The outer wrapping of the cigar is made of finest tobacco. It lends finish and aroma to the cigar.

Sizewise cigars may be classified as, Corona 9½ inches;

Petit Corona 5 inches;

Trest Petit Corona 4½ inches.

The colours are all shades of brown and do not popularly believed indicate the strength of a cigar.

Quality Pointers

1. A cigar should be smooth, firm and even to touch.
2. The open or cut end should be even.
3. A first grade cigar will produce a firm greyish ash which will hold a long time before falling.

Service and smoking of Cigars

1. Cigar should be offered in their own boxes.
2. A cigar cutter and match box must be provided.
3. If the cigar is not cut, a clean 'V' cut is suggested to allow the cigar to draw easily.

4. A cigar must only be light by match as the flame of a cigarette lighter would effect its taste.

Storing of tobacco

1. 65°F or 90°C is the best temperature for storing cigars and cigarettes.
2. A celdar wood case is preferred to any other kind of box as its allows a cigar to breath.

Some famous brands of Mild cigar are :

Coronas

Upman

Bock

Byrons

Romeoet Julieta

Cahana

Other Brands are :

Churehills

Panatellas

Margaritas

Lonsdales

Strength

1.	Colorado	C	(Strong)
2.	Colarado Claro	CC	(Medium)
3.	Claro	CCC	(Light)

Length

1.	Corona	9½ inches
2.	Corona Corona	4½ inches
3.	Half Corona	4½ inches
4.	Grown Corona	4½ inches
5.	Petit Corona	5 inches

Chapter 32

Beverages

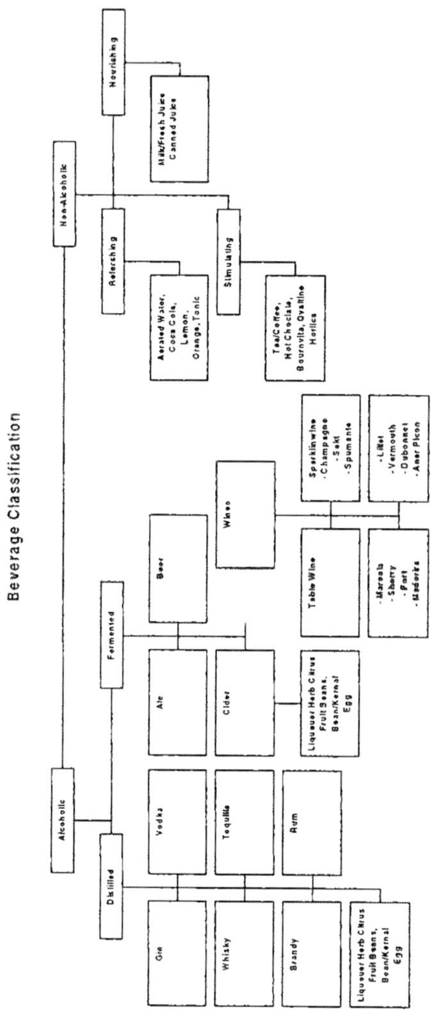

Chapter 33

Non-Alcoholic

Non-alcoholic Stimulating Beverages

Tea, Coffee.

Types of Tea

Tea is the most universally consumed of all beverages.

Green Tea

This tea is drunk plain no sugar milk and it goes through out the meal. Spiced tea is a sweetened mixture of tea, delicately seasoned with orange and lemon rings and cinnamon and cloves.

Example : Herb teas.

Orange Pekoe

Fruity in flavour, strong bodied end usually with lemon milk leaves appear straight and long.

Jasmine tea

Chinese tea (howery flavoured)

Herbal Teas

Canmolie, pepermint

Black and Green

Black tea can be drunk hot with sugar and milk or lime or lemon. It has a much stronger taste.

Souchong tea

Smokey flavoured light in body leaves appear round in shape usually drunk plain lemon, sugar.

Other teas

Masala, Ginger, Mint

Coffee

Like tea, coffee is also a universally consumed beverage. The addition of quantity of milk or cream to black coffee reduces its stimulating properties ads turns it into real food.

Filtering

Coffee experts consider the filtering of Coffee is the best method of extracting the soluble essence of ground coffee.

Coffee should be grined before being made otherwise it losses its aroma.

Coffee kept too long in the coffee urns tends to loose aroma, taste and flavour.

Cona Coffee

Cona is a brand name associated with the vacuum infusion method of coffee production. The 'Cona' has two chambers, the lower bowl is filled with worm water upto the water level indicated. The filter is placed in the upper bawl, and the require some amount of coffee grounds added. The upper bowl is then set in the lower, and the water is heated. As the water reaches boiling point, it rises into the upper chamber and infuses with the coffee grined. On reducing the heat, the upper bowl is removed and cleaned and the coffee in the lower bowl is used for service.

Decaffeinated Coffee

Coffin, a stimulant in coffee, is removed to make de-coffieinated coffee. 'Cafe Heg' in Europe of Alcoholic & 'Snaka' in America are two well known brands.

Espresso Coffee/ Cappuccino

This is an extract of coffee obtained by passing hot water and steam through tightly packed, fine grined coffee under intense pressure with the help of an espresso coffee machine. Served black it is called espresso. If milk is added and the steam applied it transform milk into a rich froth and the coffee is called cappuccino. Cinammon powder is sometime sprinkled on top.

The rules to be observed when making coffee are as follows:

1. Use freshly roasted and grined coffee.
2. Buy the correct grind for type of machine you may have in use.
3. Ensure all your equipment are clean before use.
4. Use a set measure of coffee to water 10/12 02 per gallon.
5. Add boiling water to the coffee and allow to infusion, the infusion time must be controlled according to type of coffee being used and the method of making.
6. Control temperature as to boil coffee is to spoil coffee. The coffee develops a bitter taste, due to excessive temperature.
7. Strain and serve.
8. Add milk or cream separately.
9. The best serving temperature for coffee are 82°C (180°F); and for milk 68°C (155°F).

Reasons why bad Coffee is produced (Weak coffee)

1. Water has nto reached boiling point.
2. Insufficient coffee.
3. Infusion time too short.
4. Stale or old coffee used.
5. Incorrect grinned of coffee used for equipment in operation.

Flat Coffee

1. All points for weak coffee.
2. Coffee left in urn to long before use, or kept at wrong temperature.
3. Dirty urn or equipment.
4. Water not fresh or boiled too long.
5. Coffee reheated.

Bitter Coffee

1. Too much coffee used.
2. Infusion time too long.
3. Coffee not roasted correctly.
4. Sediment remaining in storage or serving compartment.
5. Infusion at too high a temperature.
6. Coffee may have been left in urn too long before use.

Some other types of Alcoholic Coffees

Coffee	-	Benedictine is used.
Russian Coffee	-	Vodka
Jamai can Coffee	-	rum
Calypso Coffee	-	Tia - Maria
Highland Coffee	-	Scotch Whisky
Seville Coffee	-	Cointreau
Cafe Royale	-	Brandy
Irish Coffee	-	Irish Whisky

Note : *When serving coffee/tea the, waiter must remember to ask the guest if he would prefer coffee with milk or cream. Not 'black coffee' or 'white coffee' there is no such thing as white coffee or black coffee.*

Chapter 34

Grapes

There are many grape varieties found in the wine producing regions of the world but the following are the main ones :

1. Carbernet Savignon

This is considered to be one of the greatest red wine grapes in the world. It is chiefly used in combination with order bordeaux grapes, chiefly cabernet France and merlot in the production of the classic red Bordeaux wines. The grape gives the red Bordeaux wines depth, aroma Long Lasting flavour and keeping qualities. Experts say the grape can be recognised chiefly by its deep colour, the aroma almost of black currants, and by its fruity flavour. Though the basic home region of the grape is bordeaux in France, it has been successfully transplanted to make fine red wines in California, Australia, South Africa.

2. Chardonnay

This grape chiefly produces the great white Burgundies of France. It is also one of the three grape varieties used for making champagne. Some famous wines made from the chardonnay grape are Chablis and Meursault.

3. Chenin Blanc

An important grape variety used to make dry to medium sweet white wines in Loire in France, California, Australia and South Africa.

4. Gamay

This is the grape of Beaujolais. It produces red wine in France and in California, which is light in character and alcohol but with a striking fruity bouquet and a distinctive pink-purple colour. The resident large black grape of Beaujolais.

5. Gewurztraminer

The grape with a marked spicy aroma which offers an immediate and easily appreciated sensation of perfumed softness. It is Grown widely in Germany and Austria.

6. Grenache

This is a grape grown in the Southern region in the France, in California and Australia. It produces a light, fruity red wine.

7. Merlot

In both Bordeaux and California this grape is often blended with cabernet sauvignon to produce a more rounded wine. It is the dominant grape in the famous wines of Pommerol (a region in Bordeaux, France). Such as chateau Petrus. It is grown in Italy especially in the north, and in Yugoslavia.

8. Pinot Noir

This grape like cabernet sauvignon is considered to be one of the great red grapes of the world. It is used best in the classic red wines of Burgundy such as those of the cote-de-Nuits and cote de Beaune. *Pinot Noir* is also a constituent grape used in the making of champagne. The grape has a distinctive aroma.

9. Riesling

In many ways this grape is the white equivalent of cabernet Sauvignon with a normally resilient character that it retains in the many regions of the world that it has been transplanted. It is to be found probably at its best in Germany and Alsaoe (France) but the grape is also used to make some fine wines in California and Australia. Most reishing wines are dry to medium dry but in Germany in particular some of the finest sweet dessert wines are made from this grape.

10. Zinfandel

This grape is uniquely Californian, has a flavour usually described as 'Berryite' and is used for producing mainly red and some rose and lately, white wines.

11. Sauvignon Blanc

This grape is a constituent of the sweet white wines of Barsac and sauternes in Bordeaux France. Under the name of Blanc fume the grape is responsible for the fine dry wines, Pouilly fume and Scncerce in the Loire region in France. French wines called Sauvignon are often dry light white wines.

12. Trebbiano

This is the most widely planted white wine grape in Italy. It produce such wines as orvieto and Soave. In France under the name of Ugni Blanc, it produce wines of carris and as st. Emilion, it is widely grown in the cognac region.

13. Syrah

Very dark, bitter grapes used in Hermitage and Côte Rôite.

14. Cabernet Sauvignon

The chief grape in the making of claret.

15. Semillon

A delicate grape in the making of sauternes.

16. Riesleing

The clief white grape of Germany of Sauternes.

Chapter 35

Alcoholic Beverages

What is an alcoholic beverage?

Literary any potable liquid containing from ½% to 75% of Ethyl Alcohol (C_2H_2OH) by volume it is called Alcoholic Beverage. Potable alcohol is gained by :

Fermentation

Sugar in fruit or in grain developed by germination and malt is converted by the action of bacteria into alcohol. *Carbondioxide* is a by product and can be retained as in Beera nd champagne. If *Carbondioxide* escapes we hve still wines.

Sugar + Alcohol + CO_2 if there is CO_2 then we have still wines if CO_2 there then we have Beer and Champagne.

Distillation

After fermentation we heated the solution and we collect C_2H_5OH it is claled Distillation. A fermented mast of fruit is heated. Alcohol which evaporates at lower temperature than water can be trapped and condensed to liquid by cooling pure alcohol has no colour taste or semll and is used in compounding other beverages. The colour to be found in alcoholic beverages are arrived at during the aging process brought about through contact with wood, etc. over long periods or through contact with smoke or smoked wood or by stepping up the flavouring agent in the distilled spirit.

ALCOHOLIC

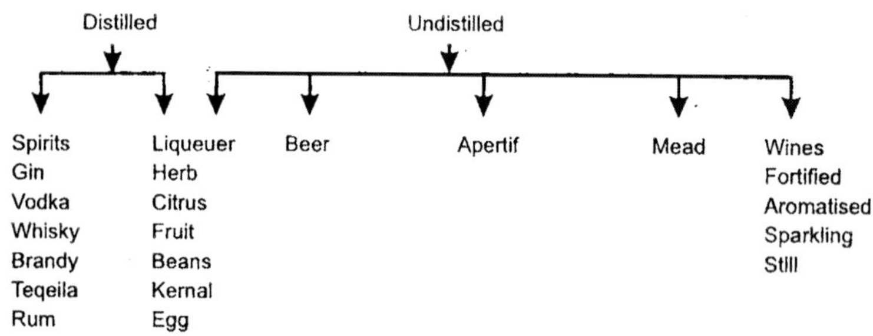
Distilled
Undistilled
Spirits
Gin
Vodka
Whisky
Brandy
Teqeila
Rum
Liqueuer
Herb
Citrus
Fruit
Beans
Kernal
Egg
Beer
Apertif
Mead
Wines
Fortified
Aromatised
Sparkling
Still

Chapter 36

Wine

Definition

Wine is an alcoholic beverage made from fermenting the juice of grapes. The stages in making wine are :

How Wine is Grown

The quality of wine depends upon the skill of the workers and the quality of the grape harvest in vintage. The quality of vintage depends upon the skill of the grower and ultimately the weather. It is easy to forget that wine growing is a horticultural operation much like any other. The annual cycle, which for the wine begins in winter with proving and taking of cutting includes all the tedious and time consuming, but vital tests that can be seen on any growing farm.

Soil cultivation, spraying, weeding, spraying again, thinning, more spraying until the times come to pick grapes that are perfect for wine making.

WINE MAKING THE BASIC PROCESS

White wine

(a) The white wine is made by red or white grapes.

(b) The Grape are first put in the crusher stemmer which breaks then and removes the stalks.

(c) The broken grapes are then fed into a cylindrical press.

(d) The press revolves and a steel screw pushes the ends together. Juice runs out into a trough and the skins are left in the press.

(e) Juice is thus pumped into the fermenting vat where its fermenting time is determined by the type of wine required.

Red Wine

Only Red (Black) grapes can be used in the making of red wine. The grapes are put into a crusher but not stemmed. After crushing the grapes, the grapes go straight into the fermenting vat with their skins - it is the skins that give the wine its colour. Fermentation continues until all sugar has been used (this takes about 2 weeks), the wine is run off without any pressing and goes straight into barrels to mature for whatever period is any necessary of the style of wine.

Rosé (Pink) Wine

Red grapes are used to make Rosé wines as before the grapes are fed into a crusher. The crushed grapes go into the fermenting vat with their skins out the process continues for a few days just as it does for red wine.

Once the wine has taken a certain point of colour, it is ran off into another vat leaving the skins behind. It then continues its fermentation in the usual way.

Pink wines are usually dry but they can be sweet.

Sweet Wine

Can be made by stopping the fermentation before all the sugar is converted into alcohol. Usually this is done by adding sulphur or filtering out the yeasts.

Dry Wine

This is achieved by allowing fermentation to continue until all the sugar has been converted.

Marc

This is the remains of skins, pips and stalk after pressing leftover them. It can be distilled to make cheap brandy, but is more frequently used as a fertilizer.

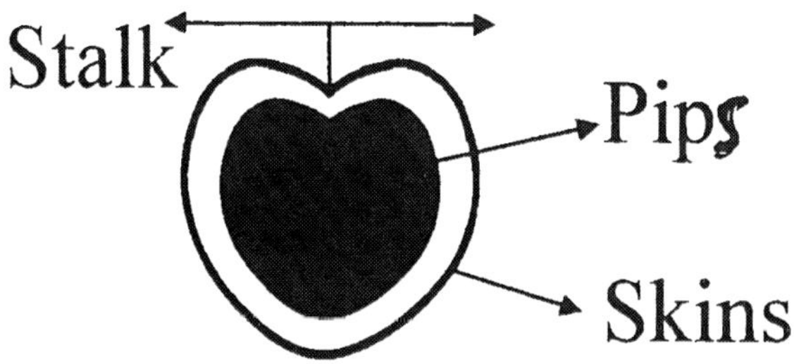

Figure

CRUSHING

Where the grapes are crushed after removing their stalks. This liquid is called Must. Pressing is the next stage where the juice is separated from the solid grape reduces (skins, pips, stem, seed). Fermentation, clarification, ageing and bottling are the remaining stages in the making of wine. White Wine is made either from white or red grapes. When it is made from red grapes the must is pressed quickly before the colour and tannin from the skin call leak into the juice. Red Wine is left as must for a time to enable the juice to colour. Rose wine is pressed a few hours after crushing, so that the juice picks up a pink colour.

During **fermentation** yeast acts on the sugar of the juice and converts it into alcohol and carbon dioxide. This process, if left alone, continues until one of three things happen :

1. All the sugar is consumed by the yeast to give very dry wine.
2. The yeast has not been able to consume all the sugar and has in fact been killed as the alcohol content in the new wine has reached its peak proportion. This results in the sugar still remaining in the new to make it sweet.

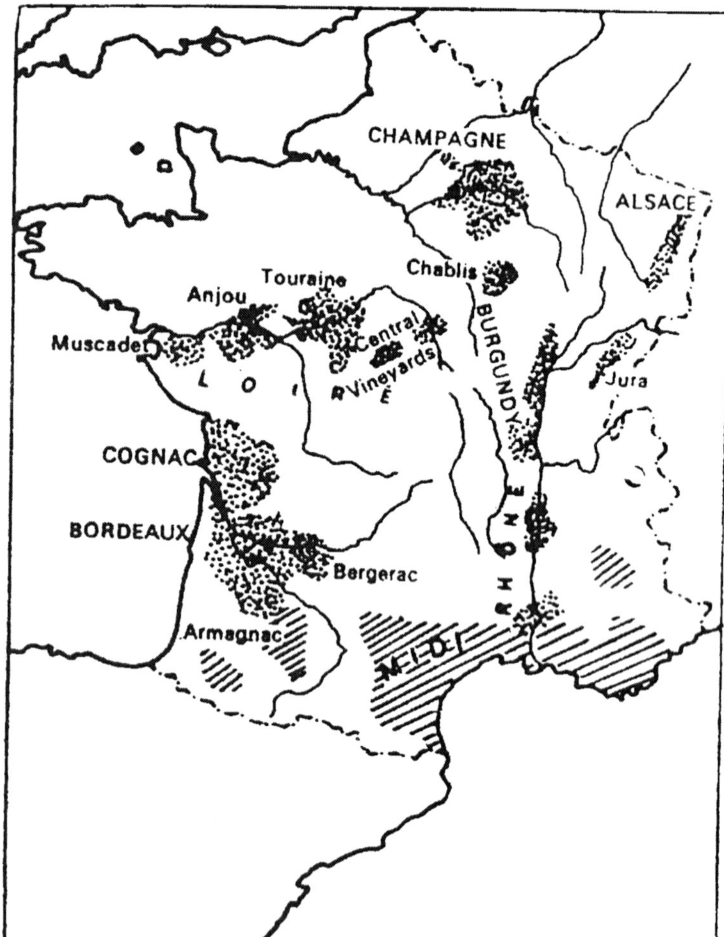

Wine growing area in France

Figure

3. More alcohol is added to the wine to discontinue the fermentation process at a particular stage as in the case of sparkling wines where a second fermentation is induced in the bottle itself.

Classification of wines

Wines are classified into the following categories :

1. Table wines
2. Sparkling wines
3. Fortified wines (Process is called fortification)
4. Aromatized wines

They may be dry or sweet (depending on the sugar content) and except sparkling wines which are only white or rose they can be red, white or rose (colour).

Table wines (Natural or still wines)

These include red, white or rose wines and are referred to as still wines as they lack carbonation. They may be dry or Sweet. The alcoholic content of these wines ranges between 14 to 16 per cent.

Example :

White : Chablis (Burgundy) Moserblumohen (Germany) and Chateau (Bordeaux)

Red : Chateau latour (Bordeaux), Beauijolois (Burgundy), Chateau Neuf du Pape (Rhône).

Rose : Rose 'd' anjou (Loire), Rose Tavel (Rhône), Fronton (Franck).

Sparkling Wines

Sparkling wines contain carbon dioxide which is achieved by a second fermentation in the bottle. The wine, therefore, is fizzy. Champagne is the monarch in this category. Alcoholic content is less then 14 per cent. These wines are drunk on festive occasions and throughout the meal.

Examples - Dom Perignon (French), Astispumanti (Italian), Sekt (German), Espumoso (Spain).

Fortified Wines

These are wines fortified or strengthened by the addition of alcohol either during or after fermentation. This increases the alcoholic content from 15 per cent to 21 per cent. These wines are drunk before or after meal.

Examples : Sherry (spin) - usually taken before a meal - Port taken after meal—Mederia, Marsala, Malaga, Vormouth

How Sherry is made

The grapes, which are always white are picked as they would be in any wine producing region and taken to the press house where they are pressed and put into the fermentation vessels. Fermentation is unique in that it is very tumultuous at first. It is said that as much as 98% of the alcohol is converted in the first 48 to 50 hours. The fermentation continues for 50 days at a much slower and quieter rate.

The new young wine is then sent to the bodegas* and the start of the maturation cycle begins. The young wines are lightly fortified and during the early stage of their maturation a unique phenomenon takes place on some of the wines a yeast - like substance called FLOR settles on the top of the wine in the casks; only these wines will develop into the fino manzanilla and Amontillado Sherries. The wines on which flor does not grow will develop from the early rayas stage into the dorosos and some of them will become cream sherry. The flor and non-flor wines are classified into their separate groups and these are fed into what is called the Solera System.

The Solera System

The purpose of this system is to achieve an absolutely constant method of blending. On that ensures that the shipper

---

* Bodegas are large church-like werehouses, the equivalent to a cellar but above ground.

** The wine of the year.

can always produce a bottle of sherry identical to the last. The method is lengthy and labour intensive, there being many stages to go through. The new wine, called the Anada** is added to a wine of the same type but one stage more mature. The process continues throughout the stages of maturation so that each cask is being topped up by the next oldest. Ultimately a mature wine of the designated style is drawn from the oldest barrel. Even then, further blending is required to achieve the necessary style required by the market. Very few unblended sherries, known as straight soleras, are even sold and they are very expensive. The reason for all this blending is that sherry is naturally the driest drink on earth and in unmodified form would find few takers.

Even straight soleras tend to be too dry for most tastes and so nearly all of the proprietary blends which are sold by wine merchants and super markets have had sweetening and colouring wines added in the final stages.

Aromatized Wines

These are wines prepared with the addition of brandy or neutral spirits and flavoured with herbs, barks and other botanicals. They are often served in combination with other alcoholic beverages as Cocktails. The alcoholic content range from 15 to 55 per cent e.g. Vermouth (Cinazano and Martini from Italy and France).

Bitters

A second kind of aromatised wines but with a little difference. They have some spirit added. The infusion of quinine, herbs, etc. give it the bitter taste as also its name. They are of two kinds :

1. Those used for colour and flavour like Augostura bittres from Trinidad) it is Rum based.

2. Those consumed on its own like campari (Italy), it is red, flavoured with orange peel.

Wine trade terms

1. Age or Ageing : Maturation in wood or bottle

2.	Apertif	: French term for alcoholic apetizer
3.	Aroma	: Taste in the mouth
4.	Bouquet	: Aroma or perfume or smell in the glass or bottle
5.	Brut	: French for very dry
6.	Carbonate	: To aerate with carbonic acid (CO_2).
7.	Cask	: Wooden containers (oak wood casks are best)
8.	Chamber	: To bring the wine to Room temperature
9.	Demi sec	: Medium dry
10.	Doux	: Sweet
11.	D.O.M.	: (Deo Optimo Maximo) Latin meaning : To God most good and most great, written on Benedictine liqueur bottles only
12.	Frappe	: Iced on the bed of crused ice
13.	Liqueur	: Distilled spirit aromatised with fruits, plants and flowers.
14.	Proof	: Standard of strength, i.e. 20° proofs is equal to approximately 10 per cent of alcohol. 50° U.P. (under proof) is equal to 75 per cent approximately.
15.	Rose	: French for pink
16.	Sec	: Dry
17.	Schaumend	: German for sparkling
18.	Vine	: Vine Spanish for wine
19.	Vintage wine	: Wine made in a vintage year
20.	Cuvée	: The contents of vat, some expressions. Such as premier curée are used to denote superior wine.

21.	Demi Sec	: Not half-dry or semi dry, as many might suppose, but well on the way to being sweet.
22.	DOC	: The Italiat system of quality control.
23.	Faded	: A wine which has lots its chief characteristics, wholly or partially, is said to have faded, usually, but not necessarily, applied to old wines.
24.	Fortification	: The process of adding brandy to natural wines to stop fermentation, preserve some of the natural sugar and boost the alcohol content.
25.	Appellation Contrólee	: The fresh system of controlling and regulating wine production, applies to the top 20% of wines.
26.	Blanc de blacns	: Wine made only from white grapes, but from more than one variety.
27.	Body	: A word used by wine tasters to denote mouth filling sensations and depth.
28.	Bourgeois	: A category of Bordeaux wines applied to individual vineyards just below "Classed" growth status.
29.	Cepage	: A french word referring to grape variety.
30.	Cru	: The word means 'growth' but is usually only applied when indicating the particular classification of wine.
31.	Green	: A word used to describe the qualities of youth and freshness in wine. It can be applied to mature wines which have those characteristic.
32.	Claret	: The English word used to denote red bordeaux wine.
33.	Climals	: The term used in Burgundy to describe an individual grower's vineyard or portion of vineyards.

34.	Clos	: Strictly speaking an enclosed vineyard, not necessarily belonging to one grower.
35.	Corked	: A serious fault, not resulting from the vinification process, but from a defective cork. The defect may develop long after bottling, and manifest itself in variety of ways. The wine way be odourless and completely flat, or it may have an unpleasant 'chemical' smell.
36.	Côte	: Means 'slope' much used in France to denote areas of vineyards. Coteaux means 'hillsides'.
37.	Hock	: An German word derived from Hocheim, used to describe Rhein wines.
38.	Maderisation	: The process of oxidation which produces a brownish colour in some wines - so that they bear a resemblance to Madeira.
39.	Noble Rot	: Botrytis cinerea, which dehydrates the grape and concentrates the sugar. Essential in the making of sauternes and other naturally sweet wines (a fungus).
40.	Nose	: Closely related to 'bouquet' this applies to the immediate impression made by the wine on the nose of the taster.
41.	Oxidation	: The deterioration of wine as a result of exposure to the air. Produces some very unpleasant smells and tastes.
42.	Pétillant	: A slight sparkle, sometimes called 'prickle' which can make white wines pleasantly uplifting. It is due to a degree of fermentation in the bottle, and should not be approved in red wine.
43.	Sulphur	: Sulphur dioxide is much used in wine making to kill unwanted bacteria. There is nothing wrong with its use but it must be carefully controlled or the wine will

		smell of sulphur and irritate the palate and throat.
44.	Tannin	: A substance derived from the skin and stalks, of grapes which helps to prolong the life of the wine. All of the very substantial and long lived red wines have it.
45.	Tough	: One of the qualities of young wine, especially those which have a high tannin content. Not an unacceptable feature.
46.	VDQS	: The second level of quality control in France, accounting for about 5% of wine. (Vins of Qualite Superior).

Serve

Red table wines with red meats.

White table wines with white meats.

Rose wines with any food.

Appetizer wines with hors oeuvres and soup.

Dessert wines with dessert or coffee.

Sparkling wines are perfect for all uses. Can be served throughout the meal.

Golden rules*

Light wines should be served before the rich or full bodied wines.

Dry wines are served before sweet wines.

The white dry wine should be served before the red wine.

Do not serve dry wine with sweet foods or any foods that have a sweet sauce.

---

* Golden Home Special (P) Marshal Cavendish Ltd., 1972, London.

The Rituals

Most mixed drinks, certain spirits which are all white wines all sparkling wine. All rose wines, all beer, all cocktail consumed as aperitifs, such as, Vodka, all aperfit wines, very dry sherry and very dry medeira are served cold (8° - 12°C).

The following generally are served at room temperature; all red wines, medium and rich sherries, madeira, port marsala and all spirits when taken neat, such as whisky, Gin, Rum, Brandy, and liqueurs.

White wine : 42° - 45° F (Chilled) (5° - 7°C)
Red wine : 65° - 68°F (Room temperature) (17° - 19°C)

Serving wine (White/Rose)

Set the white/rose wine bottle (pre-chilled) into wine bucket, with ice in it. Cover it with a clean, folded napkin and bring this wine bucket placed on a stand to the right of the person who ordered the wine.

Take the bottle out of the ice (Wine Goblet) bucket and present it to the guest with the label uppermost. The host thus has an opportunity to verify the correctness of his order. This bottle

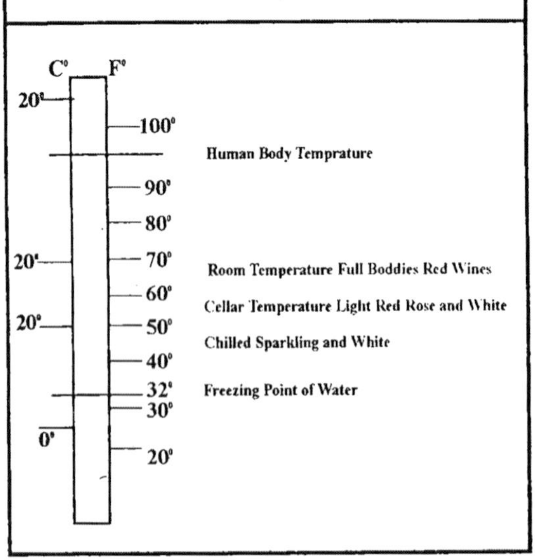

Wine Service Temperature Chart

presentation is an important part of wine service and should not be overlooked.

Cut the foil just below the bulge, which is to be found on all wine bottles. With a clean napkin wipe the mould or dirt from the neck.

Pour one ounce into the glass of the host so that he can approve the wine. Hold a waiter's cloth in the left hand when serving wine and use it to wipe the bottle, particularly when taking it out of the ice bucket.

When the host has approved the wine, pour it to the lady first (if lady on the table) then to the person sitting on the hosts right. Proceed around table counter clockwise filling the hosts glass at last. White wine glass should not be filled more than three-forth full 1/3. When pouring wine hold the bottle so that the label is always upper most and it can readily be seen. Bring the bottle to the glass on the table. Do not lift the glass in your hand because the hand warms the glass and spoils the effects of the chilled white wine. When the glass is three-forth full, twist the bottle to the distribute the last drop on the bottles rim and thus prevent it, from dripping. (You have to do with every bottle service to avoid the dropping the Drop). Always pour with the right hand from the right side of the guest. When all glasses have been served, place the white wine bottle back into the wine bucket to the host's right. You should always see that every guest's glass is replenished. It often produces the sale of the extra bottle of wine..

Rose/White Wine - 8° - 12°C chilled.

General Rules

1. Do not pour from a great height .
2. Do not rest the neck of the bottle on the rim of the glass.
3. Do not cancel the lable on the bottle.
4. Do not fill the glass to its rim.
5. Do not keep the guests glass empty for too long as temperature of the wine can change unfavourably.

Service of red wines

1. Present the wine list.
2. When the guest has ordered, place a red wine glass to the top right-hand side of the water goblet.
3. On the service guerridon assemble the following
   (a) Wine basket
   (b) Napkin
   (c) Demitasse Saucer
   (d) Waiter's cloth
   (e) Wine opener
   (f) Steak knife
   (g) Two B/B plates with napkins on it.
4. Inspect and present the bottle to the guest resting it on our lower arm boundary, a waiter's cloth folded lengthwise displaying the lable and announcing the name and year of vitage of the wine.
5. Once approved, prepare the wine basket by spreading the napkin in side it.
6. Place the bottle on the service guerridon and cut the foil, wipe the neck, unscrew and wipe again and place the wine in the wine basket or wine cradle.
7. Show the cork to the host for approval by the guest or host on the B/B Plate.
8. Pour the wine for the host for tasting.
9. After approval, serve as a detailed in the service of white wine. Remember to fill the glass 3/4 full only or upto half the monogram.

10. Find a convenient place on the table. Invert a B/B plate, cover a napkin over it and place the wine basket on it, at an incline. Remember to display the lable clearly in the basket when the bottle is empty.

Service of Sparkling wines

Open a bottle of sparkling wine by first removing the wire and foil then grasp the cork and tilt the bottle away from the guest. While holding the cork firmly, twist the bottle to allow the internal pressure to push the cork out. Keep your fingers covered over the cork to prevent it from flying off.

Some of Wine Openers (corkscrew)

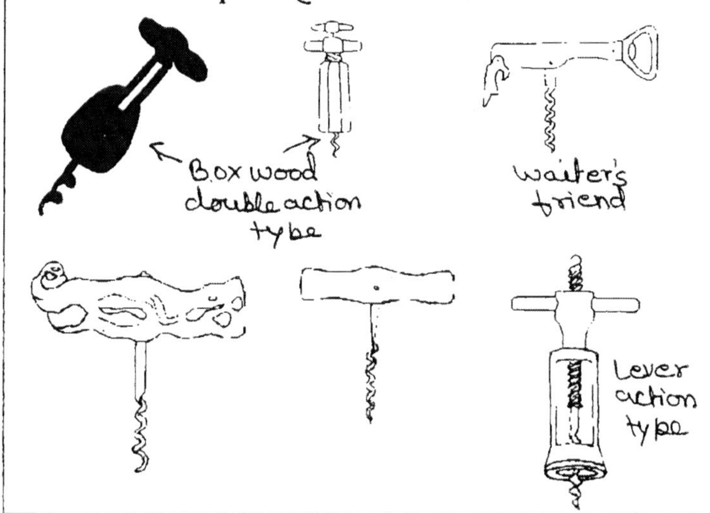

The above cover represents the placing of wine glass when the men includes the offering of white wine, red wine and Champagne.

Proceed as for white wine service.

***Note** : Sparkling Champagne is not given for taste to the host.*

Temperature = 2° - 5°C

Champagne (Sparkling wine)

Le Champagne wine takes its name from le champagne, the old province of France where it is made. Champagne, 145 Km, north east of Paris, still produces the Rolls Royece of sparkling wines.

The champagne is for happiness gaiety, laughter, for it is the joyous wine of the festival occasions.

Although there are fine grape varieties authorised for region, most champagnes are made from a mixture of two of these. The purple black 'Pinot Noir' has a colourless sugary juice that gives bady and long life to the wine whilst the small gold chardoney contributes the finesse and perfume. The champagne region is divided into three areas of production.

1. Montaygne De Reims

'Pinot Noir' wines are planted on the gentle slopes near Reims.

2. Vallee de la marne

The vineyards lie on the either side of the river Marne and grapes from here are said to contribute a particularly fine bouquet.

3. Cote de blances

This area south of the river Marne is principally devoted to growing chardoney grapes.

The production of champagne

Picking

It is critical that the grapes are picked at the correct time of ripeness. Immense care must be taken in the handling of grapes to avoid damaging them.

Pressing

The more rapidly the grapes are pressed, the less danger there is of the skins of the red grapes colouring the juice. They are put into a wooden press known as 'Marc' holds 4000 kilos of grapes and can promotes 3000 litres of Juice.

Debourbage

It is essential that all impurities in the must, be given time to fall to the bottom of the cask, so that the clean must may be drown off.

First fermentation

For about two weeks the must bubbles violently as yeasts begin converting the sugars to alcohol. From then on, if a fine bouquet and red finesse to be obtained, the fermentation has to be slow, regularly and complete.

Topping up

Casks or vats must be frequently topped up to prevent bacteria altering the character of the wine.

Racking and finning

Two or three racks a fining with bentonite or ising glass are carried out during winter to prevent the still wine from becoming cloudy.

Preparation of the cuvee

Great skill is required when blending the still wines together. The houses maintain large reserves of previous vintages to give character and flavour to wine. The wine from different vineyards and of different grape varieties must be added in varying proportions to achieve the desired balance of flavour.

Liqueur De tirage

A small amount of sugar and old wine is added along with special yeasts of a type that will fall easily. This ensures that every bottle will undergo a secondary fermentation.

Sealing

The bottles are sealed by hands with corks and agrafeelips or mechanically with crown croks wire muzzle.

Secondary fermentation

This is best carried out at 10-12° and will take 3-6 months. An inferior champagne fermented at 18°C would take 8 days. The yeasts turn the sugar into alcohol and carbondioxide which dissolves into the wine thus creating the desired stream of bubbles known as (mousse).

Ageing

The best wines are aged for 3-5 years during this time the alcohol reacts with the acids in the wine to produce the sweet smelling esters which give fine champagne is exquisite bouquet is a prolonged rest period results in mellow well balanced wine.

Remuage

The bottles are placed in a rack with oval holes, their necks being tilted slightly downwards. The specified, worker, known as the 'R emueur' holds the bottom of the bottle and give it a slight shake turn and tilt. This happens to each bottle every three days for about six weeks and at the end of this time the bottles will be standing vertically with neck downwards. The purpose of this is to drive the sediment.

Resting

Bottles are stocked upside down in the celler, with the neck of one resting in the bottom of another. The champagne can remain like this for many years, continuing its slow ageing process.

Degorgement

The neck of the bottle is dipped in a freezing brine solution, so forming a small block of ice which contains the deposit inside the neck. The pressure built up in the wine expels the ice when the cap is removed.

Dosage

The 'Liqueur de' expedition is wine of the same blend with a mixture of cane sugar and old wine added to it. A small proportion of high-class grape brandy may be included to stop further fermentation.

Recorking

A compound cork of three sections is wired down to resist the pressure of carbondioxide contained in the wine.

Shaking

Bottles are given a good shake up either by hand or machine to distribute the dosage evenly.

Resting

The wine is rested for 4-6 months to allow it to marry and settle down.

Packaging

The bottles are cleaned, labelled and despatched. This method of making champagne is known as "methode champenoise"

Styles of champagine

Brut of nature

Only a minute amount of sweetning is added in the dosage to remove the astringency of complete dryness.

Extra sec	Dry Champagne
Sec	Medium sweet Champagne
Demi sec	Sweet Champagne
Doux	Exceptionally sweet Champagne.

Matching French wine with food

Wines	*Short Description*	*Food best suited*
Graves, Muscadet, Pouilly-Fume, Chablis, Macon, Alsace, Sylavaner, Sancerre, Loire White	White Dry and Light Bodied	Sea Food, shell-fish Goat's Cheese
Alsace Riesling, White Burgundy, White Jura, Cotes-De-Provence White, Cots Du Rhone White	White, Dry and Full Bodied	Fried Fish, Cold Chicken
Anjon Rose, Jura Rose, Cotes De Provence Ros e, Cotes Du Rohne Ros e, Ros e Burgundy And Macon Medium Dry Ros e Bordeaux Rose, Corbiers Rose	Ros'e Medium Dry	Horsd oeuvre Cold Meats Entrees Roast Chicken, Gruye're Cheese
Sauternes, Barsac, Coteaux Du Layon	White Full Bodied and Sweet	Fish in Sauce Foie Gras, Desserts
Beaujolais, Loire-Red Macon Red, Bordeaux Rouge, Medoc, St. Julien, Margaux, St. Estephe, Cotes De Beaune, Cotes Du Rhone Red, Pauillac, Pomeril, Carbieres and Minervois, Bergeral Red	Red, Dry and Light Bodied	Grilled Meats, Roast Meats, Camembert, Brie Delicatessen Meats, Egg dishes
Cotes De Nuits, St. Emilion, Pomerol, Hermitage Red, Chateauneuf Du Pape, Cotes de Beaune Such as Pommard or Corton, Jura Red, Provence Red, Cote Rotie, Fitou and Cahors	Red, Dry and Full Bodied	Game Poultry in Sauce Red Meat in Sauce Strong and Sharp Cheese.
Champange		Goes with every dish Through out the meal.

Chapter 37

Alcohol Strength

1. To convert British proof to Gay Lussac (°GL)

   British proof × 4 divided by 7 = Gay Lussac or percentage of alcohol

   British proof × 4/7 = % alcohol
2. Convert Gay Lussac to British proof

   Gay Lussac × 7 divided by 4 = British proof
3. Convert British proof to U.S.A. proof

   British proof × 8 divided by 7 = U.S.A. proof
4. Covert Gay Lussac to U.S. proof

   Gay Lussac × 2 = U.S. Proof
5. Convert U.S. proof to Gay Lussac

   U.S. proof divided by 2 = Gay Lussac

Proof System

British - 100° = 57% alcohol

U.S.A. 200° = 100% alcohol

Metric System (Gay Lussac GL)

40% GL means 40% alcohol by volume

Proof

Measure of Alcohol content, alcohol applied to Gun Powder touched with flame. If ignited then 100° Proof or you can Say 57% by Volume.

Chapter 38

Whisky

Whisky is a spirit obtained by the distillation of a fermented mash of grain (usually barley, maize, rice or all together) and is aged in wood. The main countries producing it are Scotland, Ireland, Canada and the USA. Its Gaelic name the meaning water of life in become 'whisky' it must be noted that the Scottish water plays an important part in giving Scotch its special taste. Two types of whiskies are necessary to produce scotch. The first is malt whisky which gives it body and character. This is blended with grain whisky which is used for lightness. The barley is spread on the concrete floor to germinate for 8-12 days. It is then subjected to heat over fires of smokeless coal and peat. The peat gives smoke which flavours and gives an aroma to the malt during this process. The ground malt is crushed in hot water and the sweetish liquid proceed is wort. The wort is cooled and then passed in to vessels where it is fermented by yeast which attacks the sugar converts it into crude alcohol called wash. The wash is distilled twice in pot-stills by heating. As alcohol has a lower boiling part than water. It becomes vapour which is then cooled and condensed into liquid or pure scotch malt whisky. This new whisky is poured into oak wooden casks where it is matured into a pleasant mellow for as long as 15 years but not less than 3 years. It loses volume by storing. After the malt whisky and grain whisky matured, they are blended and 'married' to achieve quality and blending they are again put back in casks where they are allowed 'marry' again in the cask. It takes first years to mature and after five years a good whisky should be at its peak as regards fineness quality caramel is added to give a tanny tinge to the colour. Soft water is added to reduce the alcoholic content, then the whisky is filtered and bottled.

Malt whisky - Source is barley.

Grain whisky - Source is maize.

Irish Whisky

The barley is dried and ground with a small preparation of green malt and other cereal then mashed with boiling water in a mash tun.

Canadian (Rye)

Although Canadian whisky is usually called 'rye' the main grain used is corn with some wheat and rye.

US whisky (Bourbon)

It is produced from a mash of not less than 51 per cent corn grain. It must be aged for a minimum of two years in new charred oak barrels that must not be re-used to make whisky. Distillers hold that the natural wood caramel produced by charring gives then golden lights and particular flavour of Bourbon.

Some famous brand names

Scotch	-	Royal Salute, Chivas Regal
Whisky	-	J&B, White Horse, John Haig
U.S. Whisky	-	Old Grand Dad
or		
Bourbon	-	I.W. Harper
(Canadian or Ray)	-	Canadian Club
Irish	-	Old Bushmills

Chapter 39

Brandy/Cognac

Brandy is spirit made by fermenting and distilling of grape juice. When made from fruit other than grapes the word 'Brandy' is linked with the name of that fruit, e.g. Cherry Brandy, Apricot Brandy. When the label reads 'Brandy' alone without any qualifying description it must be derived from grape wine.

France was the first country to produce brandy for widespread commercial sale brandy is made all over France but the country is particularly known for two examples Cognac, the most famous brandy in the world, andArmagnac, its oldest brandy.

Cognac

All cognac are brandy, but all brandy are not Cognac. The only brandy in the world that may be labelled cognac comes from the charente and charente-Maritime regions in West Central France, comprising about 2,50,000 acres. The town of Cognac is in centre and lends its name to the brandy produced in the region.

How Cognac is made

The growing area of the cognac region is subdivided into 6 zones, based on soil and climatic conditions. The grand champagne and petite champagne are considered the best cognac producing districts. The use of the word "Champagne" on the label of a cognac bottle has nothing to do with the sparking wine, it comes from the word "Champ" meaning field.

Ten or so different white grapes can be legally used for cognac but the ugni Blanc also known as the st. Emillion is the major variety.

All Cognac is distilled in pot stills. It goes through two distillations. The first result in a spirit with approx. 28 per cent alcoholic content. The heart of the second distillation emerges from the still at 140 degree U.S. proof (the first and last batch) of distillate is kept a side to be redistilled. This like all newly made spirit is colourless and raw. It acquires smoothness fineness and colour as it ages in wood casks made of oak.

The age of the casks contribute to the style of Cognac, desired, new oak imports more taste to the spirit than other aged wood. A young Cognac usually is put in new oak casks but within a year it is transferred to older casks to mature more slowly. Firms marketing the lighter more delicate and softer Cognac always do their aging in used barrels. As the spirits mature they mellow. Taking on bouquet, body sublety and flavour. Although a new batch of Cognac is distilled every year there is not vintage Cognac bottled today. The Cognac sold today is a blend of different years and usually different growing districts.

Reading a Cognac Label

1. VSOP - Very Superior Old Pale : Pale does not mean any thing now but was originally intended to suggest that the Cognac natural rather than darkened on flavoured with additives.

2. VSEP - Very Superior Extra Pale.

3. XO - Extraordinarily Old.

4. Three stars of VS (very superior) or VSP (very superior pale) indicate that the minimum age for the youngest Cognac in the blend is 5 to 8 years.

   * for two year

5. VSOP, VSEP, VSO and VO indicate minimum age of 4 years for the youngest cognac with an average age between 10 and 15 years.

6. Cognac labelled with special names --napoleon, codron Blue, Triomphe, XO, extra Anniversary have a minimum age of 6 years but in fact run an average of 20 years.

7. When Cognac is labelled Napolaen it does not mean the cognac come from Napoleon's or dates back to the time of Napoleon. The name merely pays tribute to general Napoleon Bonaparte who was supposed to be very found of drinking cognac.

VSOP Brands of Cognac

1. Courvoier
2. Hennessey
3. Martell
4. Remy Martin

Chapter 40

Armagnac

Like cognac all armagnac is brandy but all brandy is not armagnac. The name armagnac can only be given to brandy made in the Armagnac area in the heart of South Western France. This area comprises three subregions Bas Armagnac, Tenareze and Haute Armagnac.

Armagnac is the world's oldest brandy, it has been made for over 500 years predating Cognac by almost two centuries. However, the total production of Armagnac is less than 10 per cent of Cognac.

How armagnac is made

Only white grapes may be used for making armagnac, the principal varieties being Ugni Blanc, Colombard and Folle Blanche.

Traditionally the wines go through a single distillation in a continuous, still and trickle out at 110 degree U.S. proof, retaining a high proportion of aromatic flavouring elements. This new armagnac is colourless with a strong bouquet and flavour that demands, longer aging for platability than is needed for cognac.

However, recent change in armagnac regulations now permit double distillations in pot stills. This yields a lighter smoother brandy which when blended with young traditionally distilled armagnaces makes them platable at an earlier age. All barrels used for aging Armagnac are made from oak wood.

Armagnac when young is more pungent in flavour than cognac. As it ages it becomes rounder, mellower and more balanced but the spirit remains full bodied with a strong fruity quality.

Armagnac labels

Armagnac used much of the same system of stars, letters and names on labels as cognac to indicate levels of age and quality and is subject to the same regulations. A minor difference is that for the Napoleon category of Armagnac. Five years is the minimum age for the youngest brandy in the blend, whereas cognac requires six years.

Some Armagnac Brands are :

1. Chabot sempe
2. Janneau
3. De Montal

Chapter 41

Gin

The spirit originated in Holland over 300 years ago when a doctor successfully combined the Juniper berry and alcohol to produce a cheap remedy for kidney complaints. The word 'GIN' is a corruption of French word for juniper 'genievre'. Gin is just alcohol, made from any source of sugar, tasteless, until redistilled with Juniper berries, angelica root, calama, coriander seed, cardamon seed and aris powder. The grain formula consists of 75 per cent corn, 15 per cent barley malt and 10 per cent other grains and the resulting spirit has to be mixed with distilled water as it is too strong to drink. Each distiller has his own secret formula which, in some cases, has not altered since the first distiller made gin.

The spirit was brought to England and acquired sufficient popularity in places such as London, Portsmouth, Plymouth and Bristol. Gin production was given full encouragement when William III from Holland and his English consort ruled England. Since then gin consumption accelerated till well after the Industrial Revolution when in 1870 Mr. Gladstone passed a bill to cut down the Victorian Gin places which were focal social points. His bill was overruled by the House of Commons and has come to stay as popular social drink.

How gin is made

Today's gin starts of with high proof spirits usually distilled from grain. They are then redistilled with or over juniper berries, citrus peels cassia bark, angelice root, anise, coriander seeds and other botanicals like vodka. Gin needs no ageing.

Once the gin has been reduced to bottling proof it is ready to be sold.

Beyond the basic juniper flavour gins have a herby/ flowery/woody taste reflecting the different botanicals used. Each distillery has its own closely guarded formula for the types and proportions of flavourings in its gin. As a rule, british gins are fragrant intense and full flavoured. U.S. gins are lighter and crisper with a muted juniper flavour and more citrus accents. They are also lower in proof : most are 80° while British gins tend to be over 90°.

Chapter 42

Clear Spirits/ Vodka

All spirits are clear when they run off the still. Colour comes later by extraction from cask during aging by the addition of caramel or by the use of wood aged. However, certain spirits are kept as clear or colourless and are not aged. The major categories of clear spirits are gin, aquavit and Vodka.

Vodka and Aquavit are primarily associated with northern Europe. The rigorous by cold climate of north eastern Europe and Scandinavia (Poland Sweden, Finland, etc.) may have been a major factor in these spirits not being aged. It was just too cold to wait around for the spirits to mellow in wooden barrels. Gin on the other hand started mainly as a drink for the poor, who could not afford the luxury of aged spirits. Improved production methods and quality controls for all clear spirits have allowed them to develop the singular character for which each is appreciated today.

Vodka

Vodka is essentially a neutral spirits which is universally thought to be made from potatoes, although it can be distilled from just about anything fermentable including potatoes. Most vodkas are made from gain.

How Vodka is made

There are differences from country to country and from producer to produce in the variety of grain used to make vodka, but it is always the one most available and economical.

The grain is made into a mash fermented and then put through a continuous still for distillation, the universal method for making vodka. The distillate comes off around 190° US proof, almost entirely free of the by-products and impurifies that contribute distinctive flavour elements to whiskies.

The new spirit may then be run though a filtering substance such as charcod or be minimally rectified. After being reduced with water to 80° US proof, it is bottled without ageing.

American Vodka

Simrnoff (the leading brand smirnoff de Czar, popor, reiska walfschwidth).

Russian Vodka

Stolichnays and Moskovskaya are the main exported brands. These vodka have smooth almost silky body and faint sweetness.

Polisy Vodka

Wybonawa Wodka ('W' is pronounced 'V') is distilled from grain and bottled at both 80° and 100° US proof.

Chapter 43

Rum

Rum is a spirit made by fermenting and distilling some form of sugar cane, usually molasses. It is believed that the word 'Rum is probably derived from Saochanma which means sugar or sweet.

The production of rum begins with the harvesting of the cane. The freshly cut cane is brought to the sugar mills, where it is passed through enormous, very heavy crushing roller that extract the juice. The juice is boiled to concentrate the sugar and evaporate the water. Then it is clarified. The result is a heavy thick syrup.

This is pumped into high-speed centrifugal machines, whirling at over 2,200 revolution per minute, where the sugar in the syrup is crystallized and separated from the other solids. After the sugar is removed, what remains? It is called MOLASSES. Sometimes this will still retain up to as much as 5 per cent sugar. The only economical way to recover or shall we say not to lose the residue of sugar is to ferment this molasses. It is taken to the distillery adding water and yeast to these molasses forms the wash to ferment, the yeast selected and the time taken governs the final character of rum.

Rums today are classified into two main groups :

1. Light bodied
2. Full bodied

Light Bodied Rum

Made in column stills they tend to be crisp and dry with a subtle flavour and aroma. Most of them have only a hint of molasses character. Some rums even approach Vodka is neutrality.

Full Bodied Rum

Made in pot stills, they have a distinct sugar/molasses character. Though they are after called sweet they are infact not, although they are aromatic and round and full in the mouth.

There is some tendency to confuse rums colour with its taste and body. It is usually thought that the lighter the colour the lighter the body and the blander the taste. This is only partly true. Colour (Usually caramel) is added to both pot and continuous still rums after distillation. Though light bodies columns still rum, do become mellow and acquire some colour, when aged in wood they do not have the potential for developing the fullness of pot still rums. However, which in a group of rums from a particular area colour will be a clue as to which are more delicate and which are heavier. The former range from clear to pale gold and the latter from amber to brown.

Peurtorican Rum

Peurtorico is the world's leading rum producing country. It sets the standard for light-bodier rum for which every aspect of production is controlled to achieve clear muted. Spirits fermentation of the molasses is carefully controlled with the use of special strains of yeast and distillation is done in column stills.

Two popular rum styles made in peurtorican are white and golden amber.

White rums

They are the dry, light and most vodka like. By law they are required to come off the still at not less than 180° US proof and can go as high as 189 degree. There are aged in wood for a minimum of one year and then filtered before bolting to remove any colour.

Bacardi is the world's most popular peurtorican white rum.

Golden or amber

Rums are bit more aromatic than the whites. They may be distilled to a minim of 175° U.S. proof and aged for a minimum of one year, Captain Morgan and Myersgolden Rich are famous brand.

All peurtorican rum are blended after aging but only with each other, never with neutral grain spirits. Although 80° U.S. proof is the standard for peurtorican rum, a certain amount of gold rum is bottled at 151°. This is cheerfully used for flambing while cooking.

Jamaican Rum

Production for classic Jamaican rum differs from the peurtorican method. The molasses is reinforced with 'dunder' which are skimmings from previous distillations and fermented with both milk and cultured yeasts. This is followed by double distillation in pot stills which yields a fairly low proof distillate. The rum are aged in Oak for 5-8 years and after blending are darkened with caramel. These rums are full bodied and richly aromatic with an enveloping tang of molasses. Much of it is aged and blended in the U.K. which is the larges customer for Jamaican rum.

Chapter 44

Tequila

Tequila is a spirit made by fermenting and distilling the juice of blue-agave plant called Mezcal. This is grown in the specified region of west central Mexico which includes the town of Tequila - after which the spirit is named.

How Tequila is made

The blue agave is carefully cultivated to produce the best quality and largest quantity of juice to make tequila, when the plants reach maturity after 8-10 years of growth. They are called PINAS (pineapples) which is what they resemble and can weight upwards of 100 pounds. After trimming the pines are halved or quartered and then baked in huge steam ovens until their starch converts to sugar. The cooked pines are shredded and crushed to extract every drop of the sweet juice or aguamiel (honey water). The aquamial is pumped into fermentation tanks and combined with cane sugar and yeast. The addition of sugar is significant. Since it can now legally constitute up to 49 per cent of the mixture to be fermented, the more sugar the less pungent the finished tequila will be. So distillers vary the amount according to the style they are seeking. Regulations require that tequila goes through a double distillation in pot stills and the distillate be run off a maximum of 100° U.S. proof. Most tequila for export is now shipped in U.K. and bottled at destination at 80° U.S. proof.

Tequila types

Tequila is available in 3 styles :

1. White or silver
2. Gold
3. Anejo

White or silver tequila is colourless and bottled with little or no aging which makes it a little harsh. Gold tequila get some inner used barrels, which both softens and tints it a bit, although the pale straw-gold colour is due to the addition of caramel.

Anejo or aged tequila is aged for at least a year in wood but in practice gets at least three years. It is smoother and mellower than either white or gold tequila. Tequila lables bear the letters 'DGN or NOM' as a quality seal. These are the initials of the Mexican Government bureaus concerned with standards.

Tequila brands

Pepe Hopez, Jose cuervo

How to drink tequila

Traditionally, tequila is taken neat, preceded by a lick of salt and followed by a bite into a lime wedge. Alternatively heat tequila can be followed by sangrita, a well seasoned tomato and orange mixture. The most popular tequila cocktails is tequila sunrise.

Chapter 45

Liqueurs

Liqueurs or cordial, as they are called in America, originated and developed through ages as a curative alcoholic herbal drinks. They are also commonly called digestive, particularly in France, where a herb based liqueur is often served after dinner to aid digestion. The name liqueur is derived from the latin word 'Liquefacere' meaning 'to melt' or to dissolve which is the way liqueurs acquire their flavour. Liqueurs are usually of high alcoholic strength and should only be taken after a meal as a digestive drink.

Definition

Liqueurs are sweetened, perfumed flavoured digestive and often coloured alcoholic beverages.

Classification

There are five broad categories of liqueurs namely :

1. *Herb liqueur* : There are flavoured with aromatic herbs, e.g. Benedictine and chartreuse.
2. *Citrus liqueur* : Flavoured mostly from peels of citrus fruits like orange and lemon, e.g., curaca, cointreau.
3. *Fruit liqueur* : Some are flavoured with almost any kind of fruits other than citrus fruits, e.g., calvados and mirebelle.

4. *Bean/Kernel Liqueurs* : Some are flavoured with bean/kernel, e.g. coffee beans, cocoa beans and fruit kernel like kahlua Tiamaria

5. *Egg liqueurs* : Flavoured with egg yolk, e.g., Advocaat.

Production of Liqueurs

Three basic methods used to produce or obtain the flavour for liqueurs are as follows :

1. *Infusion* : It is like making tea. Mostly fruit liqueurs are made by this method. Fruits are some dried fruits are steeped with spirit of average proof (120° + 130°) for approx. 6-8 months to absorb colour and aroma from fruit. The spirit is then drawn off. The mash of fruits remaining is also distilled and added back to spirit and whole is sweetened.

2. *Percolation* : Mostly plant liqueurs are made by this method. This spirit is heated and is percolated through the flavouring agents (roots, plants and herbs and beans) to absorb flavour then the sooked flavouring agents are distilled and distilled added to Percolated spirits, then the whole is sweetened and coloured.

3. *Distillation* : It is carried out in medium size pot still, in which macerated (softened by soaking) flavouring agents (roots, plants, herbs etc.) are put with spirit and is distilled the distillate which is colourless is then coloured and sweetened to become liqueur.

Making of liqueur flavour

Having got his essential components, it is time for the procedure to marry them together. An important part of this is obviously the extraction of the elements over a period of time and the clarification and stabilisation of the liqueurs prior to sale. Nobody wants the various elements to separate themselves in the bottle and fall to the bottom as sediment.

There are two methods of adding the flavour to liqueurs, known as :

1. Heat method

2. Cold method
3. Gold method

Heat Method

The heat method is best for ingredients which are rich essential oils, such as the herbs, roots peels etc.

Cold Method

The cold method is best for those substances which do not require the application of heat at boiling point to yield up their flavour and perfume, particularly the soft fruits and certain resionous. Substances which do not reach favourably to excessive heat.

Gold Method

Gold method of infusion or maceration either method may be used simply or a combination of both may be used, depending of the type of raw material used and the secret of compounding the liqueur.

Service of liqueur

Liqueur are normally served after a meal preferably with coffee. They may also be served at other times or in other styles, i.e. on the rocks with cream or as a liqueur based cocktail.

Liqueur glass in shape and size accordingly to the preference of the establishment, but have a line cut into glass to mark the measure.

Service

1. *From bar* : Liqueurs are usually served into the glasses and carried to table on a silver salver with a serviette on it. The liqueur glass is then placed at the right side of the coffee cup but not near the edge of the table.
2. *From liqueur trolley* : Liqueurs are normally offered from a liqueur trolley at the table. The wine waiter should present the trolley immediately after the sweet course is finishes to ensure that any liqueurs required will be on the table by the time coffee is served.

Uses of Liqueurs

3. The liqueur bottles are taken to the table by trolley presented and the measure poured into the glass in front of the guest.
4. If a liqueur is requested with cream then cream is slowly poured over the back of a teaspoon to settle on top of the selected liqueur without mixing it.
5. If a liqueur is requested to be served 'Frappe' then it is served on crushed ice.

A large glass used.

Fill the glass 2/3rd full with crushed ice and pour over the measure of selected liqueur.

Place two short drinking straws in the glass and served.

Liqueurs being sweet, potent and with important herbal oils are natural digestives. For this reason they are popular as after dinner drinks. As they add smoothness, texture and palatability, they are sometimes used in making cocktails.

Margarita
Tequila and cointreau.

Frigid witch
Lime Juice, strega and crushed ice.

Liqueurs are used in desserts and chocolates as they are full of flavour. Parfait grand mariner is an instance where an orange flavoured liqueur is used in the making of an ice-cream item.

Benedictine and chartreuse
Special mention must be made of these two proprietary brands as they rank as the two oldest liqueurs in the world, D.O.M. on Benedictine is (do optima maximor) to good, most good, most greatly.

Chapter 46

The Winelist

The drawing up of a wine list is akin to menu compilation, since it involves knowledge and careful attention detail. Wine lists vary in length and, therefore, in content according to the type of licence and kind of establishment.

In a popular restaurant, the customers just want to know whether the wine or sale is red, white or rose and the price by the glass, carafe, bottle, half or quarter. In a top income establishment catering for a more discriminating clienteles not only a wide selection of wines is expected but the individual wine must be properly described with a note of the vintage, the shipper and other relevant detail. Many wine lists in famous restaurants are veritable collections pieces with extensive selections of wines of all kinds. Apart from the description of the wine, bin numbers are always quoted and this ties up with the checking ordering and control administration side.

The term of wine list is generic one. It embraces all bar stock sold, namely table and fortified wines, spirits, liqueurs beer minerals, squashes, cordialsand cigars. In popular and middle steam operations the selection of wine is usually printed on the menu. In top bracket restaurants the wine list is a separate and exclusive document printed and presented with the utmost care.

From a policy angle the wine list is a permanent affair. The contents of which seldom require altering except of course the keeping of prices and vintages up to details.

Again, the importance spelling and layout cannot be overemphasised together in the avoidance of presenting customers with soiled or thumb-marked lists. As with menus, and so with wine lists the best way to explain is by example.

Chapter 47

Cocktails

Most of the people one meets in the places where cocktails 'grow' have idea that they know the origin of the word 'cocktail'. None of them, however, agrees as to what that origin is. In the 'balance', an American Periodical of May 13, 1806, we read that cocktail is a stimulating liquor, compound of spirits of any kind sugar, water, and battered sling and is supposed to be an excellent electioneering potion. This is the earliest reference to the cocktail that has been found in print.

There are so many stories regarding the origin of the word cocktail but the most authentic and incontrovertible story of the origin of the cocktail is as follows :

Somewhere about the beginning of the last century there had been for some very considerable friction between the American army of southern states and King Axototi VIII of Mexico. Several skirmishes and one or two battles tool place, but eventually a truce was called and the king agreed to meet the American General to discuss terms of peace with him.

The place chosen for the meeting was the king's pavailion and the American general was accommodated with a set in the bench as it were next to the King himself. Before opening negotiations, his majesty asked the General, as one hear to another, if he would like a drink and the American General of course said yes. The king gave a command and in a few moments there appeared a lady of entracing and overwhelming beauty, bearing a gold cup containing a strong portion of her own brewing potion.

Immediately an awed and ominous thus feel upon the assembly. The same thought struck everyone at the same time, namely that as there was only one cup either and the king or the General would have to drink out of it first and that the other would be bound to feel insulted. The situation was growing tense when the cup bearer seemed also to realise the difficulty for with a sweet smile she bowed her head in reverse to the assembly and drank the drink herself. Everything was saved and the conference can to a satisfactory ending, but before leaving the General asked if he might know the name of the lady who had shown such tact. The king said that it was my daughter Cocktil. The general replied I would see that her name is honoured for evermoore by my army. "Cocktil" of course, became "cocktail" and there exists definite unquestionable story. After this we see the cocktail in printed form.

Main types of cocktail

There is no official classification of cocktails, custom become law and thanks to the inventiveness of its originator, a single cocktail recipe can generate a whole new family of related drinks, but cocktail is a mixture of three things :

1. Base : Spirits, Liqueruos and Liquers
2. Body : all Juices + Areated Waters + Syrup + Bitter etc.
3. Garnish : Cherry + Fruit + Vegetable

Short and long drinks

Short drinks: Short drinks are served in glass which varying in size from 90-120 ml (3-4 Oz.). Their alcoholic strength can vary from about 400, the strength of the original spirit, to 10 or 12 when diluted by additions containing little or no alcohol, for example, crustas, fixes, doises frappes, smashes, sours, straights, etc.

Long drinks : Long drinks are served in glasses varying in size from 20-330 ml (4-11 Oz.) and sometimes even larger. Composed of spirits or liqueurs mixed with soda, tonic, fruit-juice, beer, wines, etc. they are generally less alcoholic and more thirst quenching than short drinks.

Classifications

Buck : A long, thirst-quenching drink made with soda or some other fizzy liquid like tonic, ginger ale or coca cola, or with sparkling wine or champagne, gin has usually provided the alcohol in a buck, but is now made with many other spirits.

Cobber : A long drink based on wine, fortified wine or spirits, easy to make and thirst-quenching. The ingredients are poured into a glass over ice cubes, decorated with fruits in season or fresh mint leaves and served with straws. Wine clobbers are made with Bordeaux, burgundy, Rhine wines, sherry, port, etc. and spirit cobblers with rum, whisky, cognac, gin, etc.

Collins : A long drinks made in collins glass with spirits, lemon juice, soda and ice. Decorate with a slice of lemon or a cherry. Tom collins is supposed to be made with dry gin and John collins with Hollands gin.

Cooler : Long drinks mixed in a large glass (250 or 330 ml) (8-11 Oz). It should not be too sweet not too alcoholic or it will be make a food thirst quencher. It is usually made of spirit, sugar or sugar syrup and ginger ale or similar fizzy drink. The best known coolers are based on rum, gin or calvados, to which lemon juice and sugar added. Coolers are also made with red or white wine.

Crusta : A short drink served in a small stemmed glass (90-150 ml - 3-5 Oz) and resembling a sour. The crusta can be made with any spirit or liqueur but it is most often made with brandy.

Recipe : moisten the rim of a glass with lemon juice and dip in powdered sugar to frost it. In the same glass put some ice cubes. I dash of Angosture bitters, 3 dashes of Maraschino, and 1 measure of spirit. Stir well and strain into the glass. Decorate with a cherry and the peel of lemon or orange cut in a long spiral.

Cup : usually a long drink with a wine base, served on ice and decorated with pieces of fruit, cucumber slices, mint leaves etc. In the past it used to be served in a large silver bowl resembling a sports trophy, the name of this trophy was hence. It can and sometime must be prepared in advance and is after served at receptions. A cup is served in a stemmed glass holding 250 ml (8

Oz) or in a glass with handle. It is related to cold punches but it is more sophisticated in decoration and after takes longer to prepare because it is left to macarate for several hours before being served.

Daisy : A short drink based on spirit bourbon being the most popular accompanied by lemon juice to which 1 teaspoon of grenadine or raspberry syrup is added. The drink is served in a small tumbler holding 120 ml (4 oz). It is generally mixed in a cocktail shaker or an electric mixer, and poured into a small tumbler containing crushed ice. It is then topped up with soda and decorated with fresh mint, raspberries or orange.

Eggnogg : Hot or cold drinks based on egg and milk to which brandy or rum is added. It can be served in a tumbler or a stemmed glass.

Fix : A short drink with a spirit based. It is made by filling a glass holding 120 ml 4oz. with crushed ice. Juice of ¼ lemon, 1 teaspoon of pineapple syrup and 1 measure of cognac or rum, gin, anis, armagnac, whisky, calvados. Sir well with a spoon, decorated with cube of pineapple syrup and half a slice of lemon, and serve with straws.

Fizz : A drink prepared in a shaker and served in a small tumbler. Fizzes are sweeter than sours because they contains more sugar or sugar syrup, thus reducing the acidity of lemon.

Egg white is usually to a fizz, but never to a gin fizz (if it is it becomes silver fizz) not to bucks fizz which is made of champagne and orange juice and it a true fizz. A fizz can be based on any spirit but gin is the most popular. Lemon juice enhances the taste and adds a touch of bitter, and sugar, egg white and soda to town.

Flip : Originally a hot drink made with beaten eggs, hot beer, rum, powdered ginger and grated nutmeg. Nowadays it can be hot or cold made with egg yolk, spirit and dusted with grated nutmeg. The best known is cold port flip, but it can also be made with cognac, gin, rum, whisky, sherry, clavados, or even with beer as in the past (Ale flip).

Floater: In this drink base should be flated. Long drink made by pouring chilled soda into an old fashioned glass or tumbler and then covering it with a measure of brandy or liqueur poured very gently over the back of a spoon held over the glass.

Frappe: A short drink, made by pouring a liqueur over finely crushed ice in a small tubler or brandy shifter and served with short straws.

To make a peppermint frappe : 2 measures of green cream dementle into a brandy glass one-third full of finely crushed ice.

Frozen : An iced drink prepared in an electric mixer with finely crushed ice. A frozen is served with straws and should be drink slowly.

Grog : A hot drink containing spirit, rum, whisky, cognac, clavados etc. and boiling water sweetened with 1-2 teaspoons of sugar and honey and decorated with a slice of lemon stuck with couple of cloves. Cinnamon powder of stick form can also be added.

High Ball : A long drink prepared in a large drinking glass with any spirit, ice and a fizzy addition (tonic, ginger ale, coca cola, bitter lemon etc.) easy to make and popular.

Julep : The essential accompaniments is fresh mint. Crush some leaves in sugar and little water to extract the flavour added finely crushed ice, a spirit, cognac, gin, rum, whisky or white wine or champagne. Stir until the glass becomes cold and frosted. Serve with straws.

Old Fashioned : Puta sugar lump in old fashioned glass, sprinkle it with 3 dashes of Angostura, add 1 measure of rye or bourbon may be another spirit and crushed ice. Stir and add soda according to your taste, and garnish with a half slice of orange and a cherry.

Pousse - Cafe : This is a after dinner drink which consists of a series of liqueurs, syrups, spirits, etc. poured successively into a straight sided glass so that they lie on top of each other but do not mix, forming a kind of raineboy. The key to success is to put the heaviest, that is the sweetest, each must be added very carefully, it helps to pour very slowly over the ever the back of the spoon. If successfully poured, a pousse-cafe is very pretty to look at and surprising to drink.

Punch : Long cold or hot drink of which the alcohol base can be varied e.g. rum, cognac, whisky, wine, cidar. It is served in a tumbler or handled glass holding about 250 ml (8oz.) garnished with fresh fruit. Punch can be prepared in advance and served in a punch bowl.

Rickey: A medium sized drink belonging to the collins group and served in an old fashioned glass. It consists of spirit, lemon juice and soda (but no sugar).

Recipe : Put some ice cubes in the glass, pour over the juice of half a lemon and 2 measures of whisky (or gin, calvades, cognac) mix with spoon, add soda and garnish with a sliced lemon.

Sangaree : A medium sized drink like a rickey but with a base of beer, wine, port of sherry, and a dusting of grated nut meg. Usually served in old fashioned glass with straws.

Shrub : Variety of Grog to which fruit liqueur or fruit syrup is added. In the old days shrub was made by macerating fruit, in spirit, it was then heat and strained. Now liqueurs or syrups are made from blackcurrant, rasberry, strawberry, etc. are used. A measure of fruit liqueur is added to 1 measure of cognac or rum and 2 teaspoons of sugar. It is then topped up with boiling water.

Sling : A long hot or cold drink made of spirit (gin, cognac, rum, whisky, calvades, etc.) lemon juice and a little sugar syrup in a cumber, topped up with iced or hot water.

Smash : A short version of mint julep but made only with spirit and garnished with fresh mint, a slice of orange, and lemon peel cut in a spiral. Serve with short straws in an old fashioned glass.

Sour : A short drink based on lemon juice and spirits, add little or no sugar just a slice of lemon and soda, some time it is with a cherry. The best known sour is whisky sour but it can made with gin, rum, cognac etc. and also with wine, champagne etc.

Today : It is a semi-long drink based only spirit and it served hot or cold one measure of spirit (rum, whisky, cognac) and 1 teaspoon of sugar in tubler glass holding 120 ml (4 oz.) Top with cold or hot water. You can add few drops of Angostura bitters.

Daiquiris : The Daiquiris are very refreshing drink but should be drunk immediately as they tend to separate if allowed to stand, it is iced drink made with rum, lime and sugar.

Martinis : The original of name martinis is martinez but now-a-day it is very famous by the name of Martini martinez cocktail made by gin, vermouth, maraschino and bitters, the extent that there is even a recipe for naked martini, which is simply gin on the rocks.

Swizzles : The stick would be inserted in the jug and rubbed vigorously swizzles between the palms of the hands. Swizzling the ice and liqueur to create frost on the outside of the glass.

Cocktail bar

Behind the bar every thing must be ready at hand. The bar tender and his assistants must have polished manners, be correctly dressed, have a good appearance and also have a knowledge of essentials of good table service as well as cocktails, liqueurs and wine.

Bar Equipment

1. Cock screw
2. Ice cutter and shaver
3. Ice scoop
4. Ice bucket and tongs
5. Shaker
6. Mixing glass
7. Bar spoons
8. Optic Measures, 30m, 60ml
9. Fruit knife and fork

10. Cutting board
11. Straw holder and straw
12. Cherry or oliva sticks
13. Strainer and funnel
14. Wine cooler
15. Wine glasses
16. Liqueur glasses
17. Cocktail glasses
18. High ball glasses
19. Tumlders
20. Servietts and paper napkin
21. Duster
22. Ashtray, match box, Plates

Few hints for the young mixers

1. Ice is nearly always an absolute essential for any cocktail.
2. Wash the ice before using it.
3. Drinks, containing fruit juices, egg, milk, cream or any other ingredients difficult to blend, should be shaken. Never fill the shaker more than half allowing plenty of room for the shake. Shake the shaker as hard as you can. Do not just rock it.
4. Never shake a drink of effervescent variety.
5. Glass should be as cold as possible.
6. Served the drink as soon as possible after mixing.
7. Always use fresh milk, cream, eggs.
8. Always use drink as attractive as possible. Eye-appeal is most essential.

Grapepeing of mixed drinks

Mixed drinks can be roughly divided into four groups.

1. Group 1 'the aperitif' which must stimulate the appetite. It must, therefore, be appealing to the eye and have an immediate effect.

2. The cocktail, which is the largest group of all and consists of all short shaken cocktails. It should contain spirit base, fruit colouring and liquer and colouring, e.g., side car, bacardi, cocktail, pink ladyect.

3. The long drinks-which usually aerated water spirit base, e.g., collins, fizzs high balls etc.

4. The stimulant-instead of going to a doctor you put your self into the hands, of the bar tender. He will prescribe for your needs, e.g., egg noggs, flips, etc.

Base of cocktail

Cocktails consists of base, body and garnish

1. *Spirits*

a.	Whisky	-	Scotch, whisky, bourborn, rye.
b.	Brandy	-	Cognac, Indian Brandies, and other fruit brandies, e.g., apricot brandy, apple.
c.	Rum	-	Bacardi, Daiguiri and Indian Rums.
d.	Gin	-	Old Tom Gin, Dry Gins, Holland Gin, Soloc Gin.
e.	Vodka	-	White Mischief, Red Reban, Smirnoff, Absolout

2. *Liqueurs*

Caintreau, Cream de menthe, Cream decacao, Marashinc, Van-de-ham, Kumnel,Absinothe, Curacao, Cherry herring, Drambuie, Grand manier.

3. *Aperitif*

Sherry, Dubonnet, Camoari Littet.

Common ingredients for cocktail

1.	Juice	-	Lime, lemon, pineapple, orange, tomato
2.	Bitters	-	Angostura bitter, orange bitter, peach bitter.
3.	Vermouth	-	Italian and French (Sweet and Dry)
4.	Syrup	-	Gomme, Rasbberry, Golden syrup, Lemon syrup

5. Water - Mineral water, Soda water, Cola drinks, Lemonade, Gingerale
6. Ice
7. Water Plain

1ST GROUP

Name	*Recipe*	*Mode of mixing*	*Glass*
1	2	3	4
Dry martini	2/3 dry gin 1/3 dry french Vermouth 1 dish anogostura bitter Squeeze of Lemon	Stir	Martini glass
Medium martini	2/3 dry gin 1/6 Dry Vermouth 1/6 sweet Vermouth	Stir	Martini glass
Sweet martini	1/2 dry gin 1/2 sweet Italian vermouth surve with cherry	Stir	Martini glass
Dry Man Hattan	2/3 Rye whisky 1/3 Dry Vermouth 1 dash angosture bitter Add a cherry	Stir	Martini glass
Medium Manhattan	2/3 Rye whisky 1/6 Dry Vermouth 1/6 Sweet Cermouth	Stir	Martini glass
Sweet Manhattan	2/3 Rye whisky 1/3 Dry Vermouth 1 dash angosture bigtter Add a cherry	Stir	Martini glass
Dubbonet Cocktail	1/2 Dry gin 1/2 Dubonnet Add a twist of lemon	Stir	Martini glass
Thistle/Rabrey	1/2 Scotch Whisky 1/2 Italian Vermouth 1 dash Augostura bitter Add a cherry	Stir	Martini glass
Bamboo	1/2 Dry Sherry 1/2 French Vermouth Add a twist of lemon	Stir	Martini glass

2ND GROUP

Side Car	1/2 Cognac Brandy 1/4 Cointreau 1/4 Fresh lemon juice	Stir	Cocktail glass

1	2	3	4
White Lady	1/2 Dry Gin 1/4 Cointreau 1/4 Fresh lemon juice	Stir	Cocktail glass
Pink Lady	1/2 Dry Gin 1/2 White of egg 1 dash Angostura bitter	Shake hard strain	Frosted glass 2 to 3 fl. OZ
Alexander	1/2 cognac 1/4 Creme de cocao (white) 1 dash Angostura bitter	Shake	Cocktail glass
Bronx	1/2 Dry gin 1/4 Fresh orange Juice 1/6 Italian Vermouth 1/6 French Vermouth	Shake	Cocktail glass
Bacarde	1/2 Bacarde Rum 1/2 Fresh lemon juice 1 dash gerandine	Shake	Cocktail glass
Orange Blosom	1/3 Orange juice 1/3 Italian Vermouth 1 dash grenadine	Shake	Cocktail glass
Angel's Kiss	1/4 Gremede Cocao 1/4 Prenella brandy 1/4 Creme de Violette 1/4 Sweet cream	Pour Carefully so that they do not mix together	Cocktail glass
Old fashion (long drink)	1 measure Rye whisky 1 lump of sugar 2 dash Angostuta bitter	Crush sugar & bitter together add lump of ice decorate with twist of lemon & slice of orange stir	Old fashioned glass or Rolly Polly
Angel face	1/3 dry gin 1/4 apricot brandy 1/3 apple brandy	Shake	Cocktail glass
Brown Lady	1/2 Dry gin 1/4 lemon Juice 1/4 van-der-Hum	Shake	Cocktail
Zipin a sip	1/4 Rum 1/4 Brandy 1/4 Whisky 1/4 Guracao Sliced fruit	Shake	Cocktail
Grass hopper	1/3 Creme de menthe 1/3 creme de cocao 1/3 double cream	Shake	Cocktail
Screw Driver	1/3 Vodka 2/3 orange Juice Ice cubes	Stir	High Boll glass
Black Russion	1/3 Coffee Liqueur 2/3 Vodka	Stir	Rolly rolly

1	2	3	4
Pink gin	1 measure gin 2 or 3 dashes of Angostura bitter	Stir	Cocktail Glass
Bloody Mary	1/5 Vodka 1/5 lemon juice 3/5 tomato juice 1 or 2 dashes worcester Sauce, Salt pepper	Shake	Glass of 6 OZ capacity

3RD GROUP

Long Drinks :			
Tom Collins	Juice of half lemon 1/2 Tab. Spn. powdered Sugar, 1/2 gily old tom, lump or ice	Shake well and strain into a roll the tumbler with soda water.	Long tumbler 12 OZ capacity
John Collins	Same as above but, using Halland gin instead of old tom gin		
Fizzes			
Brandy fizz	Juice of 1/2 lemon 1/2 tab. Spn. powdered sugar, 1/2 gin gilt brandy	Shake in shaker half filled with broken ice, and strain into a tall glass fill up the (12 OZ. capacity) glass with soda water.	
Gin fizz	Juice of 1/2 lemon 1/2 tab. Spn. powdered sugar, 3/4 gill* of dry gin	Shake in shaker half filled with broken ice, and strain into a tall glass fill up the glass with soda water	
Golden fizz	Juice of 1/2 lemon 1/2 tab. Spn. powdered sugar, 3/4 gill of dry gin yolk of 1 egg	Shake in shaker half filled with broken ice, and strain into a tall glass fill up the glass with soda water	
Rum fizz	Same as gin fizz using rum instead of gin		
Slings			
Gin sling	1 tea spn. sugar 3/4 gill of dry gin 1 lump of ice	Dissolve sugar in little water. Add ice and gin. Serve in a	

1 Gill = 90 ml.

1	2	3	4
		tumbler of 12 OZ. capacity fill in with water or soda water	
Rum sling	As gin sling, using rum of gin		
High balls			
Brandy	3/4 gill brandy	Put ice in the tumbler add brandy, 12 OZ. capacity. Fill up with soda water or ginger sle. serve with a piece lemon peel if desired or with a slice of lemon	
High Balls			
Gim High ball			
Rum High ball			
Sherry high ball	Same as above using gin, rum, sherry, vermouth		
Vermouth high ball	Whisky respectively instead of brandy		
Whisky highball			
Coolers			
Tom Collins Brandy	Juice of half lemon 1 tea spoon of sugar syrup 1 tea spoon of brown curacao 1 gill of brandy 1/2 a tumbler broken ice	Shake well and Shake or mix well and strain into a tumbler half full of 8 OZ capacity broken ice. decorate with slice of orange or lemon	Long tumbler
Gin Cobbler	As for brandy cobbler, using gin, whisky		
Whisky cobbler	respectively instead of brandy		
Rum Cobbler	3 or 4 dashes of sugar syrup 3 or 4 dashes of brown curacao, 1 gill of rum	Shake or mix well and strain into a tumbler 8 OZ. capacity half full of broken ice. Squeeze lemon jucie on top. Decorate with slices of lemon and serve with Straw.	

1	2	3	4
Coolers			
Whisky cooler	1 or 2 dashes of sugar syrup, 2 dashes of orange bitter 1/2 gill of scotch whisky 1/2 pint of ginger ale	Stir well in a long tubbler 12 OZ. capacity serve with a slice of orange	
Rum Cooler	1 tea spoon of sugar syrup 1/4 gill of lime juice 1/2 gill of rum 1/3 pint of soda water	Stir well and serve in a long tumbler 12 OZ capacity	
Punches			
Brandy Punch	Juice of 2 oranges Juice of 6 lemons 1 cup powdered sugar 1/2 pint, curacao 1 quqrt, brandy 1 spoon full grenadine 1 bottle soda water 1 large piece of ice	Pour into punch bowl over a large piece of ice. Add soda water	
Rum Punch	1 1/2 to 2 OZ. of rum Juice of 1/2 lemon 1 tea spn. of sugar, Sugar syrup Finely crushed ice	Pour the ingredients into a 8 OZ. tall glass stir and capacity serve with a mint spring	
Planters Punch	1 OZ. fresh lime juice 2 tea spn. sugar 4 OZ. old jamican rum 2 OZ. water and ice	Mix sugar with lime Juice. Add rum, water and ice. Shake and serve in tall Glass 12 OZ. capacity	
Milk Punch	4 OZ. of brandy 3 dashes of lemon juice 1 table spoon of sugar Syrup, blance of milk	This may be prepared tall glass cold (in the shaker, 12 OZ. half filled with broken capacity) or hot by heating city. The milk serve with grated nut meg.	
Sours			
Brandy Sour	1 tea spoon full of sugar syrup, Juice of 1/2 lemon	Shake well and strain into a	

1	2	3	4
	1/2 gill of brandy	medium size glass 8 OZ capacity	
Whisky Sour			
Gin Sour	Same as brandy sour, using whisky gin, rum respectively instead of Brandy.		
Rum Sour			
Toddies (Cold)			
Brandy toddy	1 lump of sugar 1 lump of ice 1/2 gill of brandy	Put sugar in a small tumbler add just enough water to dissolve it. Add ice and brandy. Stir and serve.	
Gin toddy			
Rum toddy	Same as brandy toddy, using gin, rum whisky		
Whisky toddy	Respectively instead of brandy		
N.B. Toddy may be served hot also.			

4TH GROUP

Egg Nogg	1 egg 1 table spoon powdered sugar, 1/2 gill (1 1/2 OZ.) of any spirit desired fill the glass with milk	Shake well and strain into a .....long tumbler 12 OZ capacity nut meg on top	
Flips			
Rum flip	1 Egg, 1/2 table spoon powdered sugar 1/2 gill rum	Shake well and strain into a medium size grate a little glass 8 OZ. capacity nut meg on top	
Brandy flip Port wine flip sherry flip whisky flip	Prepared in the same way, but using brandy Port, sherry, whisky respectively instead of rum.	Shake well and strain into a great a little nut meg on top	medium size glass 8 OZ. capacity

Chapter 48

Mineral Water

Mineral waters: waters of medicinal value which have in them certain mineral salts. They may be natural or manufactured. Natural mineral water may be gaseous or still manufactured. Mineral waters, however, are always gaseous in nature.

Following are some important natural mineral waters:

Value minus water

Name	*Characteristics*	*Country*
Vichy celcstin	Alkaline	France
Vichy Hospital	Alkaline	France
Aix-les-bain	Alkaline	France
Arabella	Saline	France
A Penta	Saline	France
Malvern	Table water with or without gas.	France

Manufactured mineral water

Soda

Lemonade

Gingerale

Ginger beer

Tonic water

SERVICE OF MINERAL WATER

The service is carried out in the same manner as for any other soft drink. The bottle of chilled natural water is brought from the pantry on a salver and is served in a water goblet.

The number is set in form of the quest and the bottle of mineral water opened tilting at 45°. Any mineral water left in the bottle should be set before the guest the label the guest on the top right head side of the cover. Mineral water bottle should be opened in front of the guest only. Never open on the side station.

Chapter 49

Beer

It is an alcoholic beverage obtained by the fermentation of cereal. The process is known as brewing.

Brewing is as old as human record sumerian. Babylonienm, Asynian, Egyptian record dating back to 6000 years testify to this fact that brewing was well established at the dawn of civilization. It is very good in summer. It gives refreshing. The alcoholic content is between 3 per cent to 5 per cent only.

Material required to make beer

Malt

The best cereal for use in the production of bear is barely. This cereal goes through a process which converts it to what is termed malt.

Hops

These are specially grown for brewing.

Sugar and yeast

Findings and Primings: This is a solution of sugar and hops added to some beer at racking. The function of the primings is to develop the condition of the beer by the remaining yeast reacting with the sugar to give off carbon dioxide in the cask.

Water: 90 per cent of beer is water.

Malting

First quality barley grain is taken, moistened and kept at an even tamp, until they sprout. The germination supplies the diester which inverts the non-fermentable starch into fermentable sugar. It is then charred in kiler (a type of oven).

Mashing

Girding of malt and treating with warm water to dissolve out the sugar. This liquid is called *wort* or sweet *wort*. The mineral content of water has great effect on final product.

Hopping

Hop is added to the wort and boiled at high temperature. This supplies the preservative and imports a characteristic flavour and bitter taste *Golding fugles* are two types of hops used in the beer making.

Pitching

The mixture is cooled down immediately. Yeast is added to cause fermentation (the addition of yeast to any fermentable curbing is called pitching).

Addition of sugar

Sugar may be added if necessary and the sugar may be either in the form of plain or refined sugar, or invert sugar or glucose.

Classification: after fermentation the beer is taken to large storage tank.

Finning

This is added to classify the beer. But the modern filtration is superseding the old finning method. After filtration, the beer is kept in large storage tank at law tamp.

In case of lagger beer Co_2 gas is introduced through complex automatic machine in the bottle.

Main type of beer

1 . *Drought Beer*

Drought beer is sold in bulk in Cask. Co_2 gas is introduced at the time of dispense.

2. *Lager*

A bright clear light bodied beer which is sparkling. Brewed from malt (sometimes other cereals) hope and water. The resulting wort is fermented and lager (sorted) from ageing and sedimentation. After this it is carbonated with its own carbondioxide which is gathered and added to it. All American beer are lager type. Lager beer was introduced in America by Germans in 1840.

3. *Pilsner*

The term for light beers, the original and most famous is pilsner urquea.

4. *Stout*

Very strong malt flavoured very dark ale has a sweet taste and a strong hop character.

5. *Ale*

It is an aromatic malt and cereal brew, fuller bodied and more bitter than beer. It is fermented at a higher and the yeast temperature remains on top. It is a top fermentation brew.

6. *Porter*

A type of the Ale having a rich and heavy from every dark malt is used to give a high extract. Sweeter and less happy then regular ale, brewed like stout but not so strong.

Book Beer

A special brew of heavy beer usually somewhat darker and sweeter than regular beer which is prepared in winter for use in spring. It heralds (amounces) the spring. The season last for 6 weeks.

Service of Beer

On the demand of the guest waiter has to show the bottle to the guest before opening. When guest is alowening you to serve the beer then you served the beer in beer goblet or beer mug the beer gently down the side of the glass. In case the beer is over chilled, enough froth does not form in the glass when produced. So if the guest complained that the beer is flat (NoCo_2 gas) shake the bottle lightly, froth will come up and this proves that the beer is not flat.

Beer is a brewed and fermented. Beverage made from malted barely and other starchy cerea is flavoured with hops. Beer is a generic term embracing all malt beverage.

Chapter 50

Napkin

Napkin plays a very important part in food and beverage service you found some of Napkin fold and Beverage Service this folds well give quality of service as well as hygiene to the guest this folds give general appearance to the Restaurant or Banquet Party.

There are two kind of fold

1. Simple fold, and
2. Difficult fold.

Simple Fold

Usually simple fold are used in restaurant where napkin are used in day to day apportion but it depend on the set up of the restaurant. Simple fold that may be completed more quickly and requiring us handling by the waiter and it is more hygienic.

Difficult folds

May be used on the special occasions such as B.S.D. (Buffet, Sitting Dinner) of VIP weddings party and gala dinner (gala dinner is the theme for VIP and as well as meeting of the VIP of the country.) difficult folds requiring longer time to complete, more handling and its looking crushed when unfolded to lay across the guests lap.

Size of Napkins 21m x 21m (Should be Square)

Tea Napkins 12″ x 12″

Name of the Folds

1. Bishap Cap
2. Con Cap
3. Wave Fold
4. Hut
5. Fan
6. Book Fold

Some of Napkin Folding are follows :

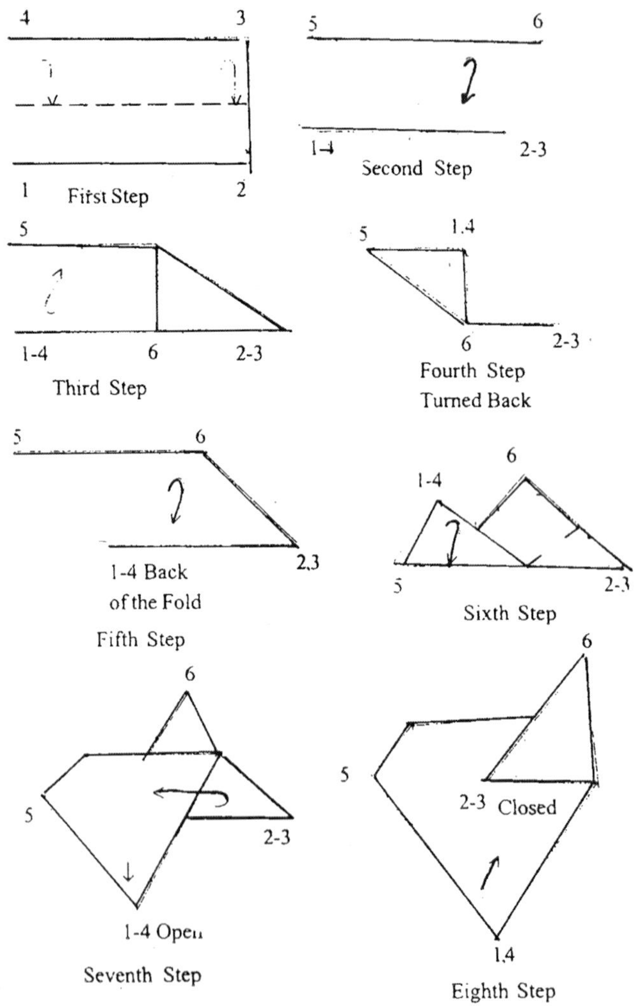

Figure

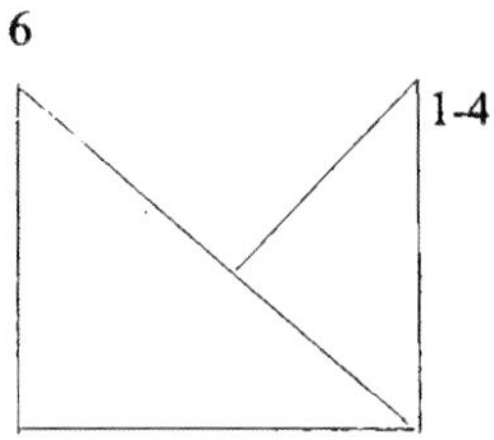

(Bishop Cap)

(Book Fold)

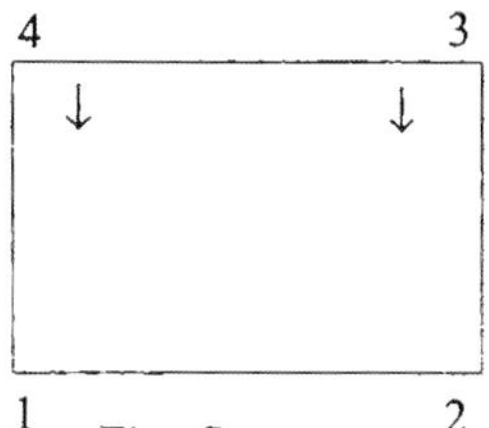

First Step

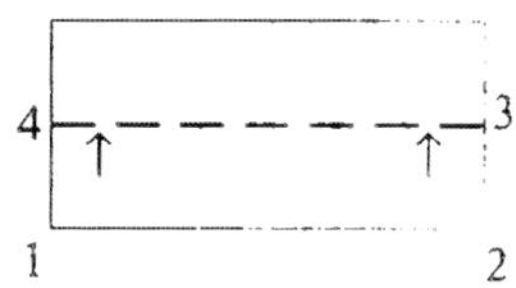

Second Step

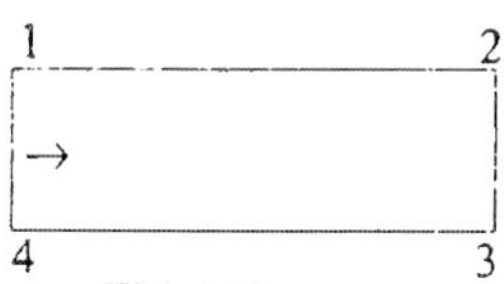

Third Step

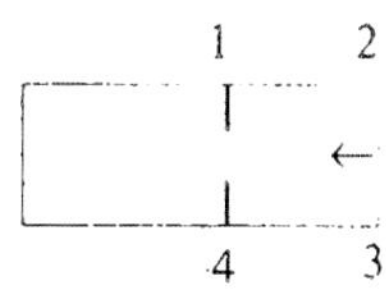

Fourth Step

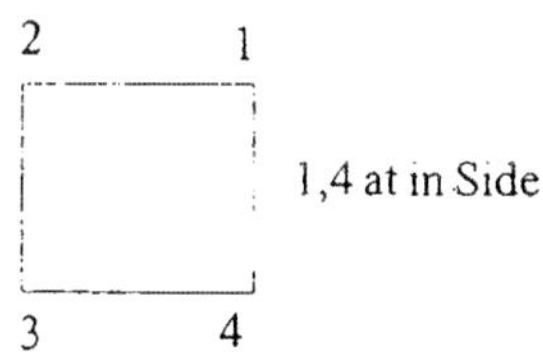

2,3 on the top

Final Step

Figure

Hut Fold

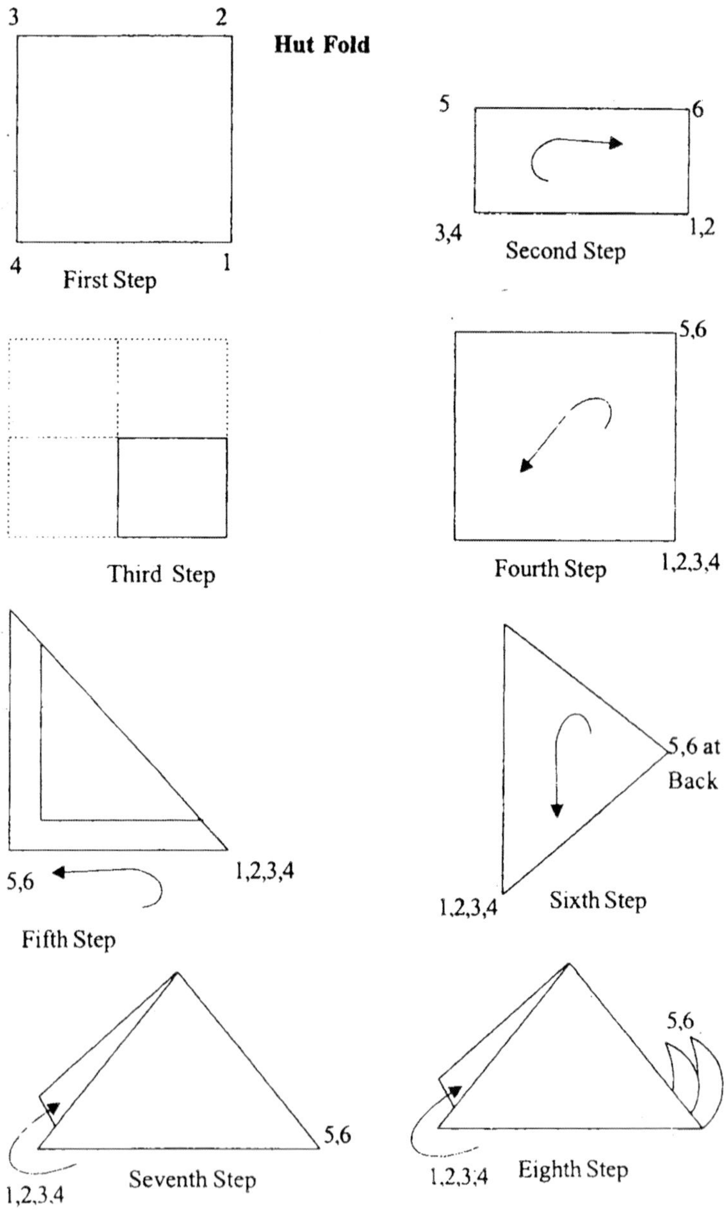

Figure

Corn Cap

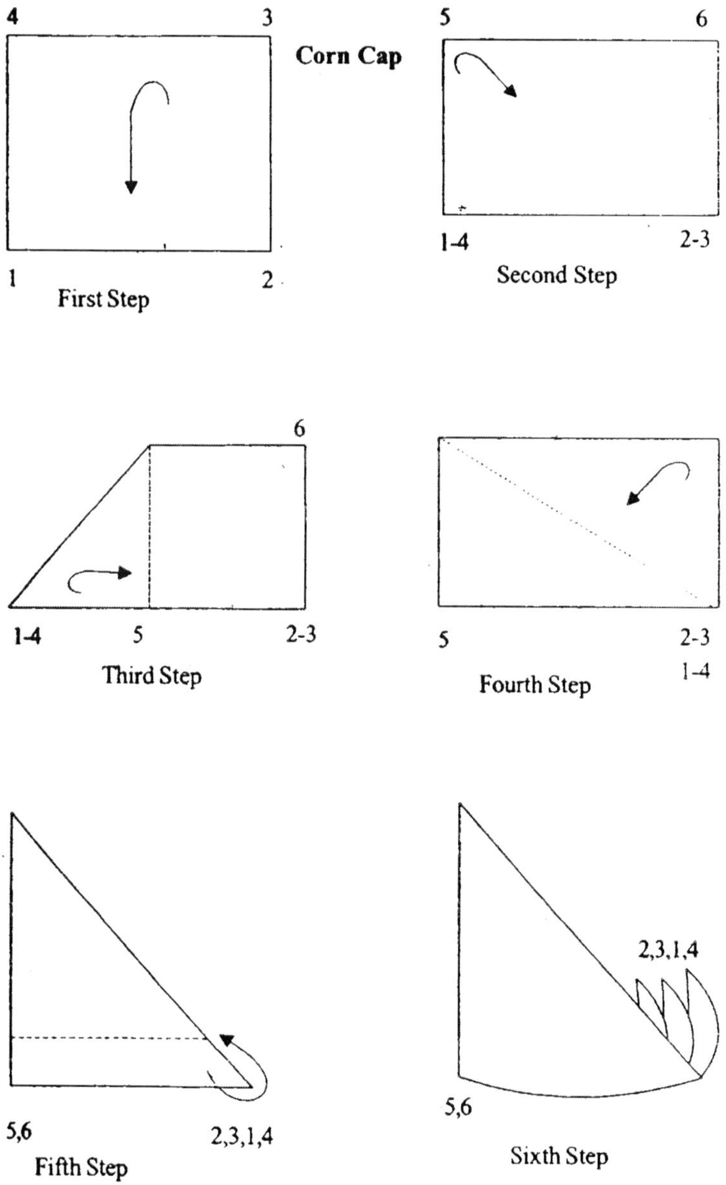

Figure

Chapter 51

Banquet

The word banquet seems to come from the word banc (bench) and, to explain how derivative, meaning a meal taken in company, was formed from the name of a seat, it has been suggested that it was on benches that the first Christians set as they celebrated their agepes in the catacombs.

Whatever the origin of the word banquet, it signifies a other Solemn and Sumptuous meal given to a large number of guests. A banquet is given on festal or ceremonial occasions, or it is given to bring together people with the same *religious* or *political* ideals, the same literary or artistic (or sometimes just gastronomical) tastes. People of the same social status or Profession, People of the same geographical or ethnic origin and so on.

People therefore gather round the same table either owing to the hecessities imposed by family life, or else to mark an important occasion and to affirm mutual ideas intellectual or sentimental.

The origins of these two reasons for gathering round a table go back to the earliest times.

On the other hand, primitive men quite naturally took their meals with their families, as it has been established that the very first man did not live by themselves, but in the groups. *Just like many animals do, such as monkeys or elephants.* These family meals consisted of rough foods that had not yet been improved by planned cooking. As soon as fire was discovered men brought meals near to it, first to dry them, then to grill and cook them.

The discovery of fire rendered the habit of gathering round the hearth to have a meal even more natural.

Archaeologists have supplied us with invaluable information about our ancestor's meals. Prehistoric strata plainly show traces of fire, heaps of ashes, in which were found bones of animals that had been used for food, and by which different prehistoric eras can be identified.

The mystic character of a banquet is also to be found in prehistory. In the origins of all the manifestations of human activity there was magic, namely the need of man to make the mysterious forces of nature favourable to him. On the walls of the *Trois Fréres caves in Ariége*, a primitive artist has painted the figure of a sorcerer in ceremonial robes, who seems to be performing a sacred dance in the middle of an immense herd. It was a sort of incantation to make hunting successful; and when the Vanquished animal lay on the ground, it was naturally divided into two part. The first for the benevolent divinities, the other for the tribe, clan or family. So men grew used to meeting to divide amongst themselves the chosen bits of the animal, and they continued to do this on the occasion of the two great events in their lives : birth and death. These were the first banquets.

Enormous progress was made in cooking when clay or bronze vessels were invented, vessels that could withstand fire and in which meats, vegetables and fish, seasoned and spiced, could be cooked.

Ever since then, large gatherings at table became possible. The era of banquets began, and the most magnificent were those that took place in the fertile lands of the orient, especially rich in spices and flavourings.

The great banquets of today - Most banquets of the past were remarkable for the prodigious number of guests who attended them and for the even more prodigious quantity of good which was served.

But at these banquets not all the dishes included in the different courses of the menu were served to all the guests.

Whilst some ate turbot or salmon, others had to make do with whiting or even cod. This also applied in the case of removes, entrées, roast meat courses and sweets.

At the great modern dinners, at which up to several hundred guests may be present (and in the twentieth century there have been banquets for 30,000 and even 50,000 persons), all the guests, if the service is correctly executed, eat the same dishes at the same time.

No doubt the number of the dishes of which a great modern dinner consists has been considerably reduced (quite rightly so) and, although the number of courses remains the same, each of these courses nowadays consists only of one, or rarely, of two dishes.

It is a rule that a great banquet, apart from the time devoted to speeches (for there are speeches at all the big dinners) does not last more than an hour and a quarter, so far as the actual serving of dishes is concerned. It would be difficult, therefore, to serve the multitude of dishes, which were an essential part in all the banquets of the past, in such a short space of time.

For some idea in the courses which it is more or less customary to include in the menu of a modern Banquet in Indian food and the general organization.

Example of Banquet Menu in India

Gala Dinner/Banquets :

State Banquets :

Chapter 52

Types of Functions in Banquet Department

I. Formal Lunches/Dinners, State Banquets

II. Reception Cocktail

High Tea

III. Buffets - Standing/sitting - Breakfast/Lunch/Dinner/ Social Functions like weddings, birthdays.

IIIa. Lunches/Dinner Meeting

IV. Conferences & Convention

(a) Workshop

(b) Seminars

(c) Training sessions

(d) Annual General Meetings

(e) Sales meets

(f) Press meets

(g) Previews

V. (a) Exhibitions

(b) Fashion Shows

(c) Dealer's Meetings/Product launching

(d) Club Meetings

(e) Musical Evenings/Dance Performances

VI.	Traditional events	- Festivals - X-Mas, Easter, New Year's Eve
VII.	Theme Banquets	
	Theme Parties	
VIII.	Outdoor Catering	

BANQUET STAFF

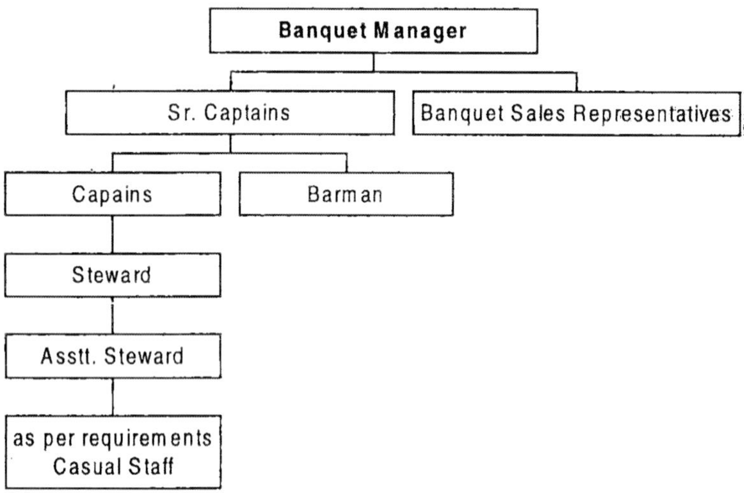

Chapter 53

Banquet Procedures (Party Booking)

Booking procedures

1. The booking may be received :

   a. over the telephone,

   b. parties calling in person, and

   c. through correspondence.

2. All the tentative bookings will be noted down in the Banqueting reservation diary along with the details specified by the party in pencil.

3. Before taking any booking, it will be ensured that there is space available on the required date and the party can be accommodated.

4. In case space is available, the booking party is requested to call in person and select manu or send a letter of confirmation and wherever possible we should call party in person.

5. A contract/function form will be filled up complete in all respect in 5 copies and will be duly signed by the booking party and Banqueting Manager.

6. Liquor will be served as per the excise rules. The booking party will be informed about them at the time of booking.

7. All the statutory laws, rules and regulations will be informed to the guest before booking the party. In case any special permission is to be taken by the guest or management, the same has to be in writing. If the management has to obtain permission for licence, a certain amount will be charged as service charges in addition to the licence fees, etc. paid to the concerned department. In case the guest is interested in obtaining his own licence, the relevant columns of form will be filled up and handed over to the guest.

   General Manager/Manager is taken. On receipt of a cancellation letter, the reservation diary will be checked and across the booking "cancelled" will be written along with the date, time, cancelled by and reference number. In case of private party, the booking party's signature will be obtained on the contract/function form to confirm the cancellation and the same procedure as above will be followed.

Amendments

Amendments regarding date of party, time, menu, of arrangements will be made in the reservation diary and a letter will be sent to the booking party to confirm the implementation of amendments or a letter of regret for not being able to make amendments as desired with reasons will be sent and all concerned departments will be notified about the amendments.

General Credit Guidelines

1. *Credit*

(a) *Film Shooting*

(i) 100% advance to be taken at the time of booking itself. This is in view of the shooting dates being cancelled quite often by film producers at the last moment, thereby resulting in the hotel not being able to sell the outlets.

(b) *Wedding/Private Dinner Functions*

(i) 50% advance of the guaranteed number of persons to be taken at the time of the booking itself, subject to a minimum of Rs.10,000/- as deposit.

(ii) Full amount of the guaranteed number of persons to be taken immediately after the function and before the guest departs. Payments from these parties should be taken in cash only.

(c) *Exhibitions*

(i) One day's rental to be taken 100% in advance at the time of booking.

(ii) For the remaining days of the exhibition if any, full amount for the following days to be taken on the day prior to the exhibition.

(d) *Conferences/Functions of Companies on Credit List*

(i) Bills to be forwarded to the Company through the Accounts Department after taking the due acknowledgments and signatures of the bills.

(ii) Parties to be informed about credit period as per the guidelines issued by the corporate finance division.

Credit cards of various agencies as well as the hotel credit cards are not valid for banquet parties.

Example of Banquet Menu

Snacks	*Non Vegetarian*	*Vegetarian*
	(Choose Any Two)	(Choose Any Two)
	Mutton Seekh Kabab	Paneer Finger
	Boti Kabab,	Paneer Tikka
	Tandoori Chicken	Vegetable Tikki
	Murgh Tikka Nawabi	Cheese Crouette
	Murgh Tikka Hariyali	Vegetable Seasame
	Murgh Malai Tikka	Seed
	Fish Amritsari	Vegetable Spring Rolls
	Fish Finger	Cashewnut Rolls
	Meat Ball	Cheese Ball
		Cocktail Samosa

Soups (Choose Any One)

Creme Chatillion (Tomato), Cream of Chicken

Cream of Vegetable, Cream of Mushroom

Sweet, Corn Soup, Creme of Doria

Salads (Choose Any Three)

Fruit Chaat, Aloo Chaat, Russian Salad, Coleslaw Salads, Hawaiian Chicken Salad, Fresh Green Salad, Kachumber Salad, Waldrof Salad, Mix Lentil Salad, Macaroni Salad

Main Course	*Non Vegetarian*	*Vegetarian*
	(Choose Any Two)	(Choose Any Two)
	Murgh Makhanwala	Mix Vegetable
	Murgh Kadai	Kesari Paneer
	Murgh Methi	Paneer Lababdar
	Murgh Korma	Paneer Makhani
	Murgh Shahjahani	Aloo Gobi
	Mutton Rogan Josh	Malai Kofta
	Rara Gosht Punjabi	Palak Kofta
	Keema Muttor	Navaratan Korma
	Chicken Saute-Hong Kong	Sarson Ka Saag
	Chicken Hot Garlic	Dum Aloo Kashmiri
	Chicken Black Bean-Sauce	Muttor Mushroom
	Sliced Chicken	Deluxe Vegetable with Almonds
	Black Mushroom with Broccoli	Mixed Vegetable (Sweet & Sour)
	Shredded Lamb in Hot Garlic sauce	Deluxe Vegetables

Dal (Choose Any One)

Dal Makhani, Dal Basanti, Rajmma Punjabi, Pindi Chana

Raita	(Choose Any One) Boondi Raita, Aloo Raita, Pineapple Raita, Kachumber Raita, Dhahi Bhala
Rice	(Choose Any One) Peas, Pulao, Jeera Pulao, Navaratan Pulao, Kashmiri Pulao, Plain Rice, Vegetable Fried Rice
Breads	(Choose Any Four) Nan, Roti, Missi Roti, Parantha, Lacha, Pudina (Parantha), Makki Ki Roti
Common	Achar, Chutney, Pappad
Desserts	(Choose Any Two) Gulab Jamun, Garam Jalebi, Kesari Kheer, Rabari, Rasmalai, Moon Dal Halwa, Ice-Cream (Strawberry, Chocolate, Vanilla) **Beverages as per Consumption** : Mineral Water, Soft Drink, Soda Water, Juices.

Sequence of Food Display of Buffets

While you are displaying the food for guest, before guest start the food you have to cheque the sequence of food. Food should be in proper sequence.

1. Soup and Soup Cup's and Plates
2. Dinner Plates with Napkins and Cutlery
3. Salad Bar with service gare
4. Rice
5. Fish
6. Chicken
7. Mutton
8. Paneer
9. Mashroom
10. Vegetables
11. Dal
12. Bread
13. Fruits
14. Desserts

Notes : Supervisor has to cheque Linen should be clean and iron and the table should be fully covered by frill.

Example of Function Prospectus

<table>
<tr><td colspan="2">Time   Day   Date   Type of Function & Occasion   Room Reserved</td></tr>
<tr><td colspan="2">Name & Address   Tel.   Guests Guaranteed
Expected</td></tr>
<tr><td rowspan="4">Menu (Food pick up at ....)</td><td>Rate per Guest : Rs.
Room Rent : Rs.
Mode of Payment</td></tr>
<tr><td>Deposit Rs.   R. No.   Date</td></tr>
<tr><td>Bar :</td></tr>
<tr><td>Arrangements & Set-Up
Exec. Housekeeper/Florist
Chief Engineer/Music
Others</td></tr>
<tr><td>Above mentioned menu, arrangements and charges are acceptable to us

Guest's Name ....
Signature ....</td><td>Banquet Manager</td></tr>
<tr><td colspan="2">Board to read ....</td></tr>
</table>

Types of Conference Setup

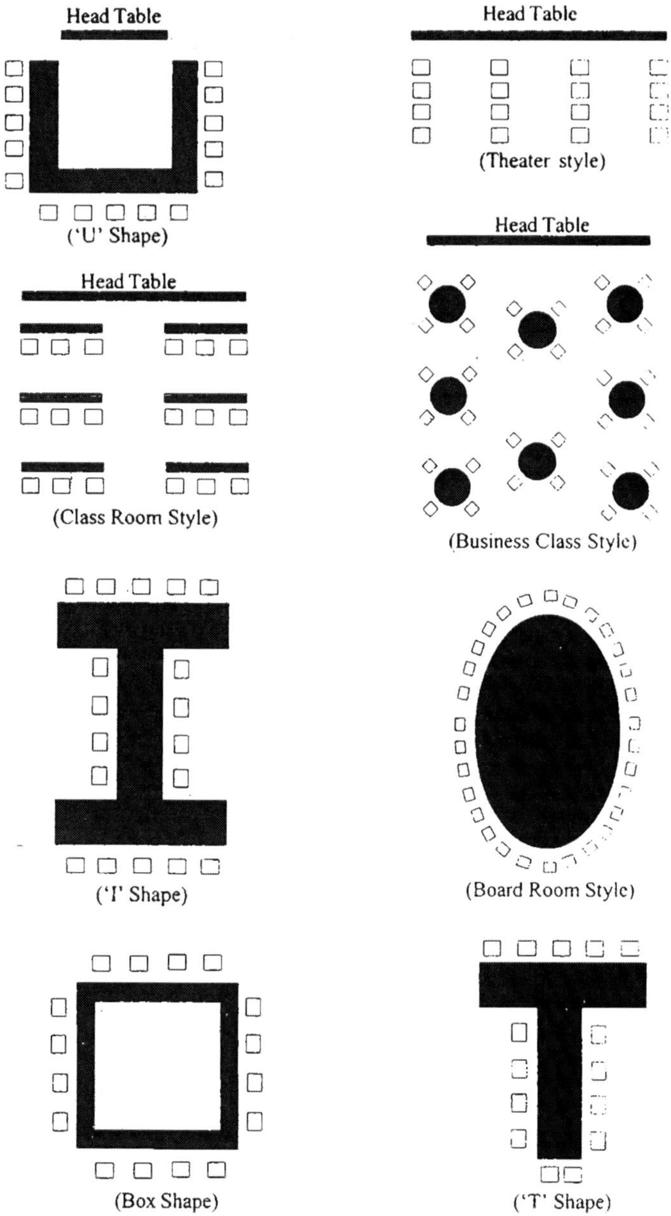

Figure

French Terminology

		French	English
Vegetables	1.	Aubergine	Brinjal
	2.	Artichaut	Artichoke
	3.	Asperges	Garlic
	4.	Ail	Garlic
	5.	Betterave	Beetroot
	6.	Carotte	Cabbage
	7.	Chou	Cauliflower
	8.	Choufleur	Cauliflower
	9.	Citron	Lemon
	10.	Celeri	Celery
	11.	Concombre	Cucumber
	12.	Champignons	Mushrooms
	13.	Epinard	Spanach
	14.	Gingembre	Ginger
	15.	Légumes	Vegetable
	16.	Laitue	Lettuce
	17.	Menthe	Mint
	18.	Oignon	Onion
	19.	Pomme de terre	Potato
	20.	Persil	Parsley
	21.	Pois	Peas

	22.	Potiron	Pumpkin
	23.	Raifor	Horse-radish
	24.	Radis	Radish
Fruits	1.	Ananas	Pineapple
	2.	Abricot	Apricot
	3.	Banane	Banana
	4.	Cerise	Cherry
	5.	Dates	Dates
	6.	Fraise	Strawberry
	7.	Fruits	Fruits
	8.	Grenade	Pomegranate
	9.	Melon	Melon
	10.	Noix	Nuts
	11.	Mangue	Mango
	12.	Orange	Orange
	13.	Pêche	Peach
	14.	Poire	Pear
	15.	Pamplemousse	Grape-fruit
	16.	Pomme	Apple
	17.	Framboise	Raspberry
	18.	Papaye	Papaya
	19.	Prune	Prune
	20.	Raisin	Grapes
	21.	Sultanes	Sultanas
Fish	1.	Anchois	Anchovy
	2.	Anguille	Bel
	3.	Blanchalle	Whitebait
	4.	Cabillaud	Cod
	5.	Eperlan	Smelt
	6.	Fletan	Halibut
	7.	Hareng	Herring
	8.	Limende	Lemon sole
	9.	Maquereau	Mackerel

	10.	Merlan	Whilting
	11.	Poisson	Fish
	12.	Plie	Plaice
	13.	Stromate	Pomfret
	14.	Saumon	Salmon
	15.	Sole	Sole
	16.	Turbot	Turbot
	17.	Truite	Trout
Shell Fish	1.	Crouletache	Shell fish
	2.	Crevettes roses	Prawns
	3.	Crevettes grises	Shrimpse
	4.	Coquilles St. Jacques	Scallops
	5.	Escargots	Snails
	6.	Huttres	Oysters
	7.	Homard	Lobster
Poultry and Game	1.	Chapon	Capon
	2.	Caneton	Duckling
	3.	Canard	Duck
	4.	Caille	Quail
	5.	Dinde	Turkey (hen)
	6.	Dindon	Turkey (cock)
	7.	Dindonneau	Young turkey
	8.	Gibier	Game
	9.	Lapin	Rabbit
	10.	Lievre	Hare
	11.	Poulet	Young male chicken
	12.	Poussin	A spring chicken
	13.	Poule	Common hen
	14.	Faisant	Pheasant
	15.	Pigeon	Pigeon
	16.	Volaille	Chicken
Meat	1.	Agneau	Lamb
	2.	Beuf	Beef

	3.	Mouton	Mutton
	4.	Porc	Pork
	5.	Poisson	Fish
	6.	Viande	Meat
	7.	Veau	Veal
	8.	Vache	Cow
Restaurant	1.	Assiette	Plate
Equipment	2.	Assiette a potage	Soup plate
	3.	Assiette de Viande	Large plate
	4.	Assiette anglaise	Large plate
	5.	Assiette à entremet	Half plate
	6.	Assiette de poisson	Half plate
	7.	Assiette à thé	Tea saucer
	8.	Assiette à café	Coffee saucer
	9.	Assiette à consommer	Consomme saucer
	10.	Beurrier	Butter dish
	11.	Buffet	Sideboard
	12.	Couteau	Knife
	13.	Couteau à dessert	Dessert knife
	14.	Couteau de table	Large knife
	15.	Couteau huitre	Oyster knife
	16.	Couteau à poisson	fish knife
	17.	Couteau à beurre	Butter knife
	18.	Couteau à bifteck	Steak knife
	19.	Cuiller	Spoon
	20.	Cuiller à entremet	Dessert spoon
	21.	Cuiller à café	Coffee spoon
	22.	Cuiller à moutarde	Mustard spoon
	23.	Cuiller à thé	Tea spoon
	24.	Cuiller à moutarde	Mustard spoon
	25.	Cuiller à service	Service spoon
	26.	Cuiller à glace	Ice-cream spoon

	27.	Ciseaux à couper les raisins	Grape scissors
	28.	Cafetiére	Coffee pot
	29.	Confituriér	Jam pot
	30.	Casse-Noisettes	Nut cracker
	31.	Demi-tasse	Cup
	32.	Huilier	Oil and vinegarcruet
	33.	Pot à thé	Tea pot
	34.	Pot à lait	Milk pot
	35.	Pince à sucre	Sugar tongs
	36.	Pince à glace	Ice tongs
	37.	Petite assiette	Quarter plate
	38.	Suciers	Saucer
	39.	Sucriers	Sugar pot
	40.	Sauciére	Sauce boat
	41.	Tasse	Cup
	42.	Verre	Glass
Misc.	1.	à la	in the style of
	2.	à la broche	roasted in a spit
	3.	au beurre	with butter
	4.	à l'huille	cooked in or served in oil
	5.	au four	backed
	6.	au gratin	gratinated
	7.	au maigre	an expression for dishes prepared without meat
	8.	à point	just done
	9.	alcool	Alcohol
	10.	Bain-marie	double boiler
	11.	banquet	a sumptuous feast
	12.	Beurre-manie	butter and flour kneaded together
	13.	Biére	Beer

14.	Bifteck	Beer steak
15.	Biscuit	literally "twice baked"
16.	Bisque	soup prepared from shell fish
17.	Beurre	Butter
18.	Beurre fondu	melted butter
19.	Beurre noir	black butter
20.	Beurre noir	nut brown butter
21.	Beurre maitre d'hotel	maitre d'hotel butter, i.e. butter with parsley
22.	Bien cuit	well cooked or well done
23.	Boisson	drink, beverage
24.	Boisson sucre	soft drink
25.	Buffet	sideboard
26.	Cacao	Cocoa
27.	Café	Coffee
28.	Canape	bread slices cut to various shapes and sizes, used plain of fried, Various tit-bits may be placed on these and served
29.	Carte du-jour	menu of the day
30.	Chaire	Flesh
31.	Choix	Choice
32.	Confiture	Jam
33.	Cotellette	Cutlet
34.	Créme	Cream
35.	Crêpe	Pancake
36.	Croute	Crust
37.	Cru	Raw

38.	Cuisse	Leg, especially of chicken
39.	Chaud	Hot
40.	Couvert	Cover
41.	Couvercle	Lid
42.	Corbeille de fruit	fruit basket
43.	Cendrier	Ashtray
44.	Chandelle	Candle
45.	Déjeuner	Lunch
46.	Dîner	Dinner
47.	Couce or Doux	Sweet
48.	Eau	Water
49.	Farce	Stuffing
50.	Farci	Stuffed
51.	Farine	Flour
52.	Frappé	On the bed of crest ice
53.	Fumé	Smoked
54.	Froid	Cold
55.	Fromage	Cheese
56.	Garcon	Waiter
57.	Garniture	Garnish
58.	Gateau	Cake
59.	Gelée	Jelly
60.	Glace	Ice, ice-cream
61.	Glacon	Ice cube
62.	Huile	Oil
63.	Lait	Milk
64.	Macedoine	Vegetables or fruits
65.	Miel	Honey
66.	Nouilles	Noodles
67.	Oeufs	Eggs
68.	Omelette	Omelet

69.	Plat du jour	special dish of the day
70.	Petit déjener	Breakfast
71.	Pain	Bread
72.	Picalitti	mixed pickles
73.	Petit pain	bread roll
74.	Pain blanc	white bread
75.	Pain Brun	brown bread
76.	Pain grillé	Toast
77.	Pâtisserie	Pastry
78.	Roohand	lamp used for gueridon cookery
79.	Rechauffe	Reheated
80.	Rhum	Rum
81.	Riz	Rice
82.	Saignant	under done
83.	souper	Supper
84.	Salade	Salad
85.	Saucisse	Sausage
86.	Sirop	Syrup
87.	Sucre	Sugar
88.	Serviette	Napkin
89.	Traiteur	Caterer
90.	Vinaigre	Vinegar

Index

C

I

K

L